Klaus Ruthenberg

Metachemistry

Klaus Ruthenberg

Metachemistry

Translation and Cartoons by John Grossman

DE GRUYTER

This volume is a translated and revised version of Klaus Ruthenberg, *Chemiephilosophie*, Berlin, Boston: De Gruyter, 2022.

ISBN 978-3-11-140002-0
e-ISBN (PDF) 978-3-11-140104-1
e-ISBN (EPUB) 978-3-11-140270-3

Library of Congress Control Number: 2024936226

Bibliographic information published by the Deutsche Nationalbibliothek
The Deutsche Nationalbibliothek lists this publication in the Deutsche Nationalbibliografie; detailed bibliographic data are available on the internet at http://dnb.dnb.de.

www.degruyter.com

To Angelika

Preface to the German Edition

Primarily inspired by reading *Dialektik ohne Dogma* by the philosophizing physical chemist Robert Havemann (1964), I decided to study philosophy seriously in the early 1980s, near the end of my degree program in chemistry. Back then I could never have imagined that in the decade that would follow I would become part of an international group whose members had developed a distinct philosophical interest in chemistry, for hitherto unclear reasons all at about the same time. Like myself, the typical members of this group, from which the International Society for the Philosophy of Chemistry (ISPC) later emerged, are chemists who have gone on to complete a second degree program in philosophy or who have studied both disciplines in parallel. Working as a chemist and university instructor naturally meant that my philosophy studies in Berlin and Göttingen extended over a longer period of time, and even this present study cannot really be said to mark their conclusion. Over the last three decades, the history and philosophy of chemistry have increasingly become my preferred field of research. Although hardly its endpoint, this study rounds off my journey to date through the fascinating field of the philosophy of chemistry. My advisor at the beginning of this philosophical journey was Lorenz Krüger, who himself had initially studied physics and then became a philosopher. He planted the seed of my conviction that the history of science and philosophy of science are best pursued as an integrated whole. Later the contact with the culturalist school of thought of the "Marburg School" around Peter Janich made a further impression on me, guiding my future development. Since my discussions with him and his colleague Nikos Psarros I have become skeptical of certain forms of so-called theory realism. Some time later Jaap van Brakel gave me the opportunity to collaborate on a research and book project on the concept of substance. As his guest at the Institute of Philosophy of the University of Leuven during several periods of residence over the span of five years, I pursued my studies of František Wald, Wilhelm Ostwald, and Friedrich Paneth in greater depth. Having since mutated into a philosophizing natural scientist myself, I have been particularly drawn to these scholars, whose gaze extended far beyond their daily work, and this study reflects that. Jaap van Brakel also taught me to take actual chemical practice seriously in philosophical terms. In recent years I had the opportunity to conduct research at the Department of History and Philosophy of Science of the University of Cambridge and was subsequently granted the honor of becoming a Life Member of College Clare Hall of that university. This was made possible by the support of Hasok Chang. He greatly advanced my philosophical development with his own works and by way of our collaboration on our "acidity project." Since the modern philosophy of chemistry began to flourish in

https://doi.org/10.1515/9783111401041-001

the 1990s, I have continually profited from Joachim Schummer's criticism. I remain greatly indebted to him and not only with respect to the present work.

In addition to the philosophers already mentioned, I would also like to express my heartfelt thanks to the following fellow travelers who to varying degrees and at different times have directly or indirectly had an influence on my work in the philosophy of chemistry: Michael Akeroyd, Davis Baird, Marina Banchetti-Robinho, Bernadette Bensaude-Vincent, Alfred Dirschbacher, Joseph Earley, Tony Edmonds, Martin Eisvogel, Apostolos Gerontas, Elena Ghibaudi, Rom Harré, Clevis Headley, Michael Heidelberger, Robin Hendry, Claus Jacob, Endla Lõhkivi, Jean-Pierre Llorred, Olimpia Lombardi, Farzad Mahootian, Juan Camilo Martínez González, Ave Mets, Paul Needham, Jennifer Rampling, Carsten Reinhardt, Guillermo Restrepo, Daniel Rothbart, Eric Scerri, Eugen Schwarz, Jens Soentgen, Pieter Thyssen, Brigitte Van Tiggelen, Sylvia Wenmackers, and Rainer Zimmermann.

My particular and abiding admiration I send to those who unfortunately are no longer among us: Joseph Earley, Rom Harré, Peter Janich, Lorenz Krüger, Elisabeth Ströker, Krishna Vemulapalli, Rein Vihalemm.

The present study was largely written during the coronavirus pandemic. It represents a slightly revised edition of the work I presented to the Faculty of Philosophy and Social Sciences of the University of Augsburg as a dissertation (title: *Was ist Chemie?*). I extend my heartfelt thanks to my advisors Jens Soentgen and Uwe Voigt, who as "midwives of coming into being" have done an amazing job in patiently showing me both the paths to follow and the paths best not taken.

This book arose from the intention of putting my standpoint in the philosophy of chemistry into a sustainable and arguable form. It would please me greatly if it would help to avoid failing to see the forest of philosophy of chemistry for the trees.

I thank my dear wife Angelika for everything.

Klaus Ruthenberg, Bad Rodach, January 2022

Preface to the English Edition

The present work is a translation of my book *Chemiephilosophie* from 2022. After giving it some thought, I decided to retain the basic structure of the original book. I felt minor additions and corrections were warranted only in a few places in the interest of better developing certain concepts and their relationships. I am fully aware that some historians will find this book too "thin," some philosophers will find it insufficiently "analytical," and chemists may even find it superfluous. I nevertheless welcome criticism especially when it is justified. The book bears the title *Metachemistry*. Those who initially take issue with this expression will have no other choice but to study the book in its entirety in order to gain a reliable impression of what I mean by it, for here as in the case of the expression *metaphysics* a brief description of two or three sentences would be at least unsatisfactory and possibly even meaningless.

On this occasion I express my particular thanks to my old friend and former fellow student in Berlin John Grossman for the translation work and for his "systematic art."

Klaus Ruthenberg, Bad Rodach, January 2024

https://doi.org/10.1515/9783111401041-002

Contents

Table of Abbreviations

CFCs	Chlorofluorocarbons
CTR	Causal theory of reference
DDT	Dichlorodiphenyltrichloroethane
DU	Dobson Units
ECD	Electron-capture detector
EDTA	Ethylene diamine tetraacetate
FDA	Food and Drug Administration
FID	Flame ionization detector
GC	Gas Chromatography
GDR	German Democratic Republic
HF	Hydrogen fluoride
HI	Hydrogen iodide
ISPC	International Society for the Philosophy of Chemistry
IUPAC	International Union of Pure and Applied Chemistry
MO	Molecular orbitals
MPG	Max Planck Society
MS	Mass spectrometry
MSD	Mass selective detector
PFTBA	Perfluorotributylamine
QTAIM	Quantum theory of atoms in molecules
SHE	Super heavy elements
TIC	Total ion current
WMO	World Meteorological Organization

https://doi.org/10.1515/9783111401041-003

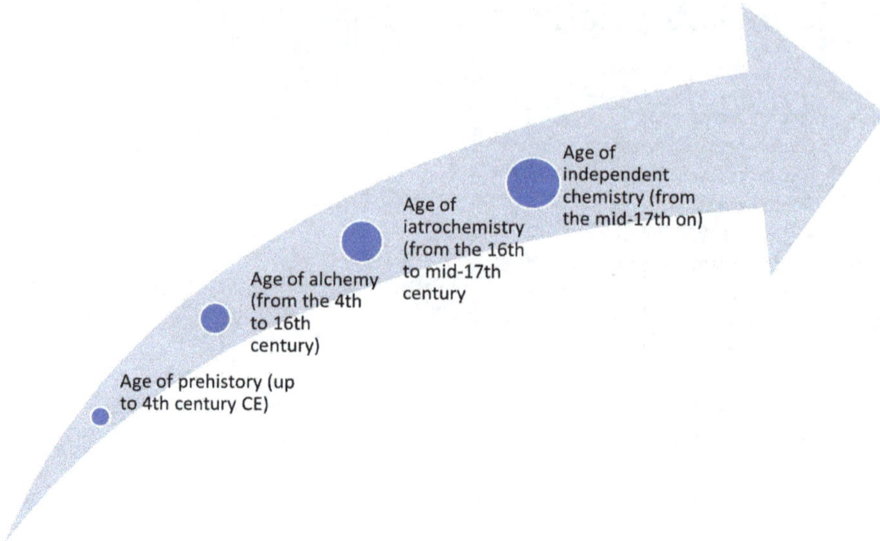

Fig. 1: The historical epochs of chemistry according to Georg Lockemann (1950). This representation is not intended to depict a continuous advancement; advancement in chemistry invariably exhibits quantitative and qualitative aspects as well as continuous and erratic aspects.

1 Introduction – Self-Descriptions and External Assessments

We encounter substance in the form of sensory experiences. This may be the feeling of air resistance when riding a bicycle, which shows us that something can be transparent and very light and yet is not nothing. This may be our encounters with liquid in the rain, when swimming, when drinking, or when excreting sweat or liquid metabolic products. This may be the many solid objects that, in contrast to air and water, persistently resist bodily penetration. Add to this our experiences of the mutability of certain things in the world in which we live: A drop of water evaporates on a hot stove; another drop of water freezes to ice on a cold windshield; and this ice melts again when we defrost the windshield. We can therefore proceed from a widespread intuitive understanding of the concept of substance that is practically independent of culture. When I use the term "substance" in this study, I assume a macroscopic entity with certain characteristics that are independent of form. These include spatial extension, divisibility, ponderability, and certain inclinations (dispositions).[1] In searching for an answer to the question of what chemistry is, it seems expedient to provide a rough preliminary historical outline of the phases of development of the scientific discipline that we now refer to as chemistry. In his brief history of chemistry, historian of chemistry Georg Lockemann (1871–1959) divided the main phases of the history of chemistry into four sections (Figure 1).

It is not possible to establish any one date as the beginning of modern chemistry. All attempts to subdivide the entire field into historical periods proceed, as does Lockemann, from certain premises. He provides this coherent observation:

> It was only very late that the realization gained acceptance that the actual task of chemistry is to research the characteristics of the various substances and their reciprocal transformations. This occurred in the 17th century. (Lockemann 1950, p. 5)[2]

Although the concept of substance itself requires clarification, this characterization initially sounds quite plausible. However, a different position is also found in the literature, one that regards the transition from the atomism of ancient times to the current microphysical interpretations as a more or less logical, quasi natural development. The controversial debates that have thus arisen define

1 Cf. the characterization of substances in Ostwald (1907, pp. 6–7). See Soentgen (1997, 2008) regarding the use of the term "inclination" (German "Neigung") in the present context.
2 Translations are throughout by the author or the translator if not otherwise noted.

https://doi.org/10.1515/9783111401041-004

the main background for this study. The central question is: “Does chemistry have its own metaphysics?” or also “Is there a metachemistry?” The background of the latter question, which is answered affirmatively by Gaston Bachelard (1884 – 1962), is the observation that the great majority of contemporary philosophers working in metaphysics are physicalists. They define the relationship between body and mind by proceeding from a materialist standpoint and therefore draw on achievements of theoretical physics that they regard as fundamental. Although the “mind-body problem” will not be addressed in this study, I would like to note that for reasons I cannot ascertain chemistry is not mentioned in the relevant discussions (cf. as an exception Earley 2008). I assume that there are basic premises and fundamental operations that, taken together, can be deemed “metachemistry” in that they are sufficiently distinct from prevailing metaphysics.[3] Thus I feel the former term is legitimate and can supplement the customary latter.

In this introduction I would initially like to provide a commented outline of previous definitions of chemistry or its precursors and then address the “stepchild issue” that has plagued chemistry from the philosophical side. I proceed from the assumption that the search for the nature of substances is an abiding characteristic that has continued throughout every epoch mentioned.[4] In the following section I will provide a historical outline of the self-descriptions in chemistry based on a selection of textbook definitions.

1.1 Self-Descriptions

How does one define a great natural science? There are obvious problems:

> It is always difficult to give beginners a precise yet comprehensible definition of the science that they desire to become familiar with, for a strict definition generally requires knowledge of a part of the phenomena with which this science concerns itself. (Regnault 1851, p. 1)

3 cf. Earley (2008, 2009) and Seifert (2023).

4 I concur here with Elisabeth Ströker who says something very important about patterns of thinking in chemistry: “These patterns of thinking […] are never plainly and simply something past: They are also reflected in the current methods of science, even stratified in their contemporary concepts.” (Ströker 1968, p. 747) The historical periods identified by Lockemann have usually been characterized very similarly even in the recent history of chemistry. Jost Weyer (2018a, 2018b) speaks of three great epochs, the chemistry of ancient times and the Middle Ages (early chemistry), the chemistry of the 16th through 18th centuries (early modern chemistry), and chemistry since the 19th century (modern chemistry).

Henri Victor Regnault (1810 – 1878), the author of these lines (and of one of the most important textbooks of his time), presents a version of the study-of-substance definition in which chemistry is represented as a natural science that concerns itself with the phenomena that cause a "complete change in the nature of the bodies." If we consider the quotation and the definition, we must ask whether this is a definition adapted to the level of knowledge of beginners or the strict definition that satisfies experts. If we assume the latter, then the author's remarks do not deviate substantially from those that were customary in the field at that time (see below). When one examines the second part of the quotation, it would seem reasonable – if one excludes a trivial literal interpretation (a useful definition of a discipline indisputably requires knowledge of these) – to supplement it as follows: The understanding of the strict definition of a science requires knowledge of those phenomena with which the science concerns itself. Then it would still remain unclear how the students can recognize which phenomena belong to the science they are studying and which do not, seeing as they are not yet familiar with this science. As paradoxical as this situation may seem, it is entirely realistic: At the beginning of their degree program, future chemists do not know what they are actually studying. This reveals a fundamental difficulty in defining an entity as complex as an entire natural science. On the one hand, the term definition demands a strict and clear, generally applicable, and non-circular delineation (primarily with respect to content, yet also with respect to method) using simple concepts; but on the other hand, the two areas are distributed among so many different individual experiences, techniques, and now also disciplines that an all-encompassing yet invariably simplifying summarization simply seems impossible.

The earliest written characterizations of chemistry (actually of its precursors) appear in the late period of alchemy, such as in the writings of Andreas Libavius (ca. 1550 – 1616). In his most important work on alchemy *Alchemia* of 1597, we read:

> Alchemy is the art of perfecting magisteries and extracting pure essences from mixtures by separating the corpus. (Libavius 1597, p. 1)

It would not be correct to translate "magisteries" as "substances" without comment, and it would be equally wrong to immediately think of transmutation, creating gold, or the philosopher's stone (although other authors mention this context). In writing this work, Libavius was not at all interested in the things that many still assume pertain to the whole of "alchemy," which indeed never existed in universal form. Rather Libavius understands "magisteries" as something like the active essence of things and substances, that is the inner "pith" of the empirical substances.[5] Except for a more precise identification of these "magisteries," the preparative operational aspect of alchemical/chemical activities is in any case made clear in this self-description. Cleaning, separation, and extraction are operations that one performs on material substrates especially for medical and pharmaceutical (iatrochemical) purposes as well as for others. Libavius feels that one could also speak of alchemy as the art of proceeding appropriately (*ars bene laborandi*) as well as of separating the pure from the impure (*segregatio puri ab impuro*).

Because of its preparative operational nature, I refer to the self-description for which Libavius formulated a precursor as the art-of-separation definition. It is found in textbooks until well into the 18th century. One example is the introductory statement by the influential phlogiston theorist Pierre Joseph Macquer (1718–1784):

> To separate from each other the various substances that form part of the composition of a body, to study particularly each of these, to detect their characteristics and similarities, to further divide these themselves where possible, to compare and combine them with other substances, to bring them together and combine them afresh in order to restore the first mixture with all its characteristics: This is the object and primary purpose of the art of separation. (Macquer 1752, p. 1)

The concept of the chemical compound[6] appears in rudimentary form or is assumed in such characterizations. Compositionism assumes that there are constituents, and these are not far removed from elements. I understand "compositionism" as that field of research in which the composition of the specimens under study is the focus of interest.[7] It is the foundation of the theoretical and methodological shift that is often referred to as the "chemical revolution." Compositionism is the program that in chemistry experienced its zenith during the course of the 18th century with Lavoisier and superseded "principlism," at least in the conventional interpretation. In the latter, chemical substances are described and classi-

5 However, Jost Weyer writes in his history of chemistry (2018a, p. 364): "The magistery is a medically efficacious substance that has been freed of its outer impurities, the 'body' arguably the inactive residue."

6 Ursula Klein (1994) argues convincingly that it was the compound concept that developed at the turn of the 17th to 18th century that crucially influenced the beginning of modern chemistry – in contrast to the widespread view that corpuscular hypotheses played an important role.

7 I take this expression from Hasok Chang (2012, pp. 37–42), who uses it to distinguish from the "principlism" of the 17th and 18th centuries. The attitude behind compositionism regards analysis primarily as (essentially cognitive) "breaking down" and is thus one main target of my argument in this study.

fied according to certain phenomenological characteristics. This system knows no elements in the modern sense. In conceptual terms, the qualitative and quantitative investigative methods here stand face to face (principles vs. composition). At about the same time, Mikhail Vasilyevich Lomonosov (1711–1765) writes:

> Definition I. Chemistry is the science of the changes that occur in the mixed body – insofar as it is mixed [...].[8] Definition II. The practical part of chemistry consists in the historical knowledge of the changes in the mixed body [...]. Definition III. The theoretical part of chemistry consists in the philosophical knowledge of the changes in the mixed body. (Lomonosov 1741, pp. 68–69)

Here Lomonosov strives for a characterization of chemistry that does without information from the current state of relevant research, which he regards as superfluous. Thus in a certain sense he wants to clarify the *what and why questions* but not the *how* questions. In terms of methodology this is legitimate for a definition, yet despite the subdivisions his approach seems poorly comprehensible and lifeless. The Dutch naturalist and physician Hermann Boerhaave (1668–1738) provides a more readily understandable version of the art-of-separation definition:

> Chemistry is an art of changing bodies with clever tools and in so doing discovering their effects and the causes of these effects. [...] Its object is every body that comes to mind; and these bodies all fall under three kinds which are known as the three kingdoms. (Boerhaave 1755, p. 1)

The three substance realms Boerhaave mentions are the minerals, the plant kingdom, and the animal kingdom.

The transition to a kind of self-description one could term a study-of-substance definition is gradual. One interesting example from the transitional phase is the following:

> [Chemistry's purposes are:] 1) the investigation of all mixing ratios and knowledge of all mixing components of natural things; 2) the ascertainment of all conditions of possible mixing changes in every mixed thing; 3) the determination, derived from experience, of all actual mixture laws and the intimation of all possible mixture phenomena; and 4) the study of

8 The text continues: "Explanation. I do not doubt that there will be many for whom this definition will appear inadequate and who [will criticize] the lack of the principles of separation, amalgamation, cleaning, and other expressions with which many who write about chemistry needlessly overload their studies yet which are included in the term 'mixed body.'" Here Lomonosov weakens the operational and preparative character of the art-of-separation definition. František Wald later demands exactly the opposite, namely emphasizing the "how" questions (see Chapter 3).

> all relationships of the chemical activities to the remaining non-chemical (cosmic and individually organic) emanations of force from beings existing in time and space. (Kastner 1814, p. 1)

Carl Wilhelm Gottlob Kastner (1783–1857), Justus Liebig's teacher, argues here far more explicitly from a compositionist perspective than the proponents of the art-of-separation definition before him.[9] About twenty years previously Jeremias Benjamin Richter (1762–1807), the pioneer of stoichiometry, had published the following statement:

> Chemistry or the art of separation is the science *) of those proportions in which certain parts of the bodies known as elements occur relative to one another. (Richter 1792, p. 1)

Richter provides an explanation for the marked passage that very succinctly characterizes the transition mentioned. In it he criticizes the view of chemistry as a simple skilled craft, that is to say as an "art," because in this way one could be a separation artist without understanding what one's own action is based on. At the same place he states:

> For several millennia chemistry was only able to bring forth processors of this sort and no others; even today there is still a large number of these same workers bearing the title chemist. It has been left to the current century to represent chemistry not only as an art but also as a science. (Richter 1792, pp.1–2)

Richter emphasizes the scientific nature of chemistry which he associates with being mathematizable. He was undoubtedly familiar with Immanuel Kant's *Metaphysical Foundations of Natural Science*, which was published in 1786. In the preface to this milestone in the philosophy of science, Kant repeatedly mentions the flourishing chemistry of his time, for example in the following sentence:

> So long, therefore, as there is still for chemical actions of matters on one another no concept to be discovered that can be constructed, that is, no law of the approach or withdrawal of the parts of matter can be specified according to which, perhaps in proportion to their densities or the like, their motions and all the consequences thereof can be made intuitive and presented a priori in space (a demand that will only with great difficulty ever be fulfilled), then chemistry can be nothing more than a systematic art or experimental doctrine, but never a proper science, because its principles are merely empirical, and allow of no a priori presentation in intuition, consequently, they do not in the least make the principles of chemical appearances conceivable with respect to their possibility, for they are not receptive to the application of mathematics. (Kant 1786/2004, pp. [470]–[471], cf. Kant 1786/1982, p. A X)

---

9 In the introductory work quoted Kastner also considers the history of chemistry in an extensive table; see the copious study by Bettina Haupt (1987).

For the purposes of discussion, this powerful sentence can be divided into several parts. The initial comment contains an indirect indication of Kant's characterization of chemistry. He mentions "chemical effects of the materials on one another" and "approach or withdrawal of the parts." These are elements of a definition of the kind we frequently find in textbooks after the fundamental renewal of chemistry. The "approach or withdrawal of the parts" belongs more to the theory of affinity than to the concepts of element or even atom. Further along in the *Metaphysical Foundations*, in the section "General Remark to Dynamics," the mechanical effect is distinguished from the chemical one:

> The action of moved bodies on one another by communication of their motion is called *mechanical*; but the action of matters is called *chemical*, insofar as they mutually change, even at rest, the combination of their parts through their inherent forces. (Kant 1786/2004, p. [530]; original author's emphases)

Kant already regards chemistry as the study of substance (if we equate his "matters" with "substances"). At the same spot he notes the difference between the separation of mixtures and the decomposition of compounds. It also becomes obvious that he dislikes (mechanical) atomism.[10] He obviously has insight into the historical development of scientific ideas and therefore leaves open the possibility that chemistry could yet progress to become a proper natural science. It can be assumed that he has followed the dispute about phlogiston. Presumably the emancipation process of the young natural school (or natural science) from alchemy on the one hand and from medicine in particular on the other has not entirely failed to impress him. This conspicuously detailed treatment of chemistry in the preface of the *Metaphysical Foundations* may be the consequence of an intensified awareness of the developing discipline.

The mention of convergence or divergence of the parts and of the proportion of the density of these parts (materials) leads to a second aspect. Kant at this point also hints at a program of research that a short time later – with the exception of Richter's stoichiometry – began to bear fruit: In 1797 Joseph Louis Proust (1754–1826) formulated the law of definite proportions, followed in 1804 by the law of multiple proportions formulated by John Dalton (1766–1844).

The pessimistic content of the parenthetical remark in Kant's long sentence is pivotal in the discussion of the status of chemistry. Here Kant expresses the opin-

10 Michael Friedman (1992) offers a detailed analysis of Kant's position with respect to the atomism of his time. See Höffe (2011) for an enlightening discussion of the Kantian classification of sciences and chemistry's position within it. Jaap van Brakel (2006) gives a very detailed study of Kant's reflections on chemistry.

ion that despite the fundamental possibility of developing into a proper science, which he does not exclude at this point, chemistry will hardly ever be able to move in this direction.[11] In his *Entwicklungsgeschichte der Chemie*, historian of chemistry Hans Eduard Fierz-David (1882–1953) states that the "fascination with the substances represents an irrational element in modern chemistry as well" and that in broad areas of chemistry mathematical and physical observations yield very little. He goes on to state:

> Yet in many cases one can later demonstrate that the observation follows mathematical laws; however, it is not yet possible today to predict all chemical reactions with calculations. In this sense we may also qualify modern chemistry as being a "systematic art." (Fierz-David 1952, pp. 2–3)

This primacy of the a posteriori mode can still claim to apply today. Kant's hope of a priori explanations within the sciences of substances is indeed unfulfillable as the latter have to do with real objects, hence theoretical abstractions – that is to say the translation into purely logical concepts – are therefore subject to "natural" limitations. Yet this does not in any way diminish the status of chemistry, rather it is typical of chemistry and thus a metachemical condition. In his very meticulous study regarding Kant's "chemistry," Martin Carrier accordingly states:

> Kant's classification of chemistry as an improper science stems in no way from a lack of respect for chemistry. In an obituary published […] in 1804 […] Ludwig Wilhelm Gilbert emphasizes Kant's chemical knowledge and his interest in chemistry. Kant's esteem for chemistry is also evident in that he [in the *Critique of Pure Reason*, B XII–XIII] places the experiments of Georg Ernst Stahl, the founder of phlogiston chemistry, next to those of Galileo and Torricelli and sees in them the paradigmatic embodiment of the scientific method. Apart from that it would be inaccurate to read Kant's term "improper science" as a euphemism for "non-science" as chemistry in the 18th century especially due to Stahl had left far behind it the status of a school of art that collects cooking recipes. (Carrier 1990, p. 194)

It should be noted that the frequently encountered interpretation of Kant's underrating of or lack of respect for chemistry as a science is unfounded. Furthermore, I feel that it is worth reflecting on whether the meaning of a *systematic art* must be evaluated as unfavorably as Carrier explicitly formulates it here and as many other experts both in the natural sciences and in philosophy, usually less explicitly, take for granted. In the following chapters I will return to this point at appropriate junctures.

---

11 This comment has such a resounding effect that Fierz-David (1952, p. 2) completely skipped over it in his otherwise very comprehensive Kant quotations.

Whether Richter actually followed Kant's guidelines consciously or more likely supported the push for restructuring made necessary by the general developments and culminating in the so-called "chemical revolution" is a question that must remain unanswered due to lack of evidence.

Let us continue with the self-descriptions in chemistry. In the 19th century the art-of-separation definition gradually changed into the study-of-substance definition. A fully developed example of this type of definition and one well specified by closer observations is provided in the statements by Leopold Gmelin (1788–1853) in the first volume of his work *Handbuch der Chemie:*

> The object of chemistry is merely those changes of the bodies that result from affinity. As these changes consist 1) in compounds of inhomogeneous materials forming a homogeneous whole and 2) in the occasionally associated precipitations of inhomogeneous substances from homogeneous wholes, chemistry can be defined as the study of the compounds of inhomogeneous bodies forming homogeneous wholes, and of precipitations of inhomogeneous substances from homogeneous wholes [...]. Chemistry is not merely a *science* but also an *art*. (Gmelin 1843, p. 1; original author's emphasis)

A series of similar descriptions of chemistry as a scientific and technical field of the qualitative changes of substances could be cited.[12] The transformation from what was once a practical ancillary discipline of medicine (from the beginning, that is from the early 17th century on, chemistry as a subject of university study was primarily iatrochemistry) to a science in its own right proceeded from the realization that it had become necessary to separate applied chemistry from pure chemistry as Johann Gottschalk Wallerius (1709–1785) programmatically demands[13] in 1751 and as many textbook authors then follow through on conceptually.[14] This paves the way for a novel definition of what is now pure chemistry and for the expansion of applied chemistry into various substance-related activities. In addition to the pharmaceutical applications, applications in metallurgy, glass making, agriculture, paint production, brewing, and other areas are now also subsumed under the terms "technical chemistry" or "economic chemistry." Added to this is the theoretical development of chemical ideas, which at the latest with Stahl's phlogiston theory in the 18th century led to the formation of a stable pan-European scientific community.

In his work *Grundzüge der Theoretischen Chemie*, first published in 1890, the co-founder of the periodic system of the elements Julius Lothar Meyer (1830–1895)

12 I again refer the reader to the historical standard study of German-language chemistry textbooks: Haupt (1987).

13 See Meinel (1984).

14 See Erxleben (1775), Wallerius (1780), and others in Haupt (1987).

formulates in a particularly compact form, albeit with a thorough introduction, this version that has remained widely accepted to this day: "Chemistry is thus the science of substance and its changes." (Meyer 1893, p. 2) As the modern element concept gains acceptance, self-descriptions begin to appear which emphasize the chemical elements, that is substances that cannot be further divided into other simple substances. A modern example of this version of the study-of-substance definition comes from an introductory university textbook of inorganic chemistry:

> Chemistry is a natural science. It concerns itself with the study of the human environment accessible either directly via the senses or indirectly via suitable instruments [...]. Chemistry is thus the science that concerns itself with the possible combinations of the 104 known elements. (Schmidt 1967, p. 1)

A concept of substance is not mentioned here but is presumably assumed. If one also disregards the fact that the environment accessible to the senses or to instruments also includes light, sound, heat, and even electricity, then the "possible combinations of the 104 known elements" is the point of this self-description that is particularly worthy of criticism. Do these 104 elements (more than a dozen have since been added ...) belong to this environment accessible to the senses or to instruments? And if yes, then do not our knowledge and skills of initially representing these elements, one could also say the *systematic art* of doing so, belong to chemistry? How do we know about these substances that we call elements? In the further course of this study, especially in the chapters on Wald and Ostwald, I will more closely examine questions of this sort.

There are further specifications and thus qualifications of the study-of-substance definition such as the following one:

> The object of chemistry is the study of substance characteristics and changes in chemical material. Chemical material is understood to be material of the kind that can be constructed directly from atoms and the ions and molecules derivable from them. (Baumgärtel et al. 1972, p. 1)

Here, too, the definition speaks from the perspective of the result aspect, not to say from that of chemists' jargon, and it also remains unclear, ignoring for a moment the circularity of the expression "chemical material," how one comes to atoms and how this specific material (whatever material may be) "is made up" of atoms and the other particles. An interdisciplinary team of authors from former East Germany arrives at a similar definition.[15] The authors emphasize the difference between

15 This concerns the chemist Karlheinz Richter and the philosopher Hubert Laitko. The latter was

the object and content of a scientific discipline. In another article, Laitko, one of these authors, states that the object definition belongs to the content and is its shortest, concentrated expression in undeveloped form and must therefore be deduced from it (Laitko 1967, p. 113). Definitions describing content are to be systematically excluded in this view. This team of authors rejects the study-of-substance definition as too broad and maintains that "the object of a science" is defined "by the fundamental laws intrinsic to all phenomena of the field" (Richter & Laitko 1962, p. 1282). Following a detailed discussion of various aspects, the authors propose their own version of a definition of chemistry:

> The object of chemistry is the formation of compounds from the elements and the reciprocal transformation of the compounds. These processes do not destroy the unity of the atoms and are always based on the association of atoms to molecules and the dissociation of molecules into atoms. The elementary reaction is the starting point of chemistry; typical of the chemical form of motion are macro-processes that are based on the elementary reactions and closely linked to them, but that cannot be reduced to them. (Richter & Laitko 1962, p. 1293)

This, as mentioned previously, is the modern atom-centric definition of chemistry. Several aspects of this suggested definition appear to me to be problematic or unacceptable. First, the substances we have been calling elements for over two centuries are themselves the object of chemical activities; in other words, one does not simply encounter them in the environment as they (or at least some of them) stand in the cabinets of a collection of chemicals. From a methodological standpoint, the knowledge of what an element is should not be simply assumed in such a definition. Furthermore, compounds are mentioned, but not (directly) their dissociation. The term "compound" requires explanation. The expression "molecule" as the smallest unit of a chemical compound requires it as well and is also poorly chosen (although it is understandable that one would seek a generic term for what can also be called "formula unit"). The problems with the concepts of "atom" and "unity" are more serious. Apart from the fact that the experts in the science of substances only experiment with atoms in exceptional cases, are a chlorine atom and a chloride ion unitary or only potentially unitary? Thus we may conclude that it is unavoidable but difficult to deduce the object of a science from its current content. As we see, the use of multiple technical terms from the current theoretical inventory that are thought to increase precision does not necessarily lead to an improvement in the definition.[16]

---

his country's leading philosopher of chemistry; cf. Laitko (1967, 1996) as well as Laitko & Sprung (1970). See also my discussion of the significance of quantum chemistry in the present study.

16 I feel it is nonetheless correct that the version of the definition discussed here emphasizes the manufacturing and process character of chemistry. See Chapter 2.

In chemistry it is not a matter of achieving a high degree of mathematization or physicalization, although both need not be detrimental, rather it is a matter of controlling the substance environment, both the natural one and the man-made one. As a result, I prefer the conventional study-of-substance definition, for example as formulated by Wilhelm Ostwald:

> As the quantity and external shape of a body belong to its arbitrary characteristics that can be changed at will, one observes in chemistry the bodies without regard to quantity and form. Bodies that are observed in such a sense with respect to their specific characteristics are known as substances. It is thus the substances that form the object of chemistry. Those substances that have matching specific characteristics are said to be identical. (Ostwald 1907, pp. 6–7)

Whereas the modernist talk of subatomic particles crops up sporadically in some 20th-century chemistry definitions,[17] this has not caused the object of chemistry to undergo any fundamental transformation within the scientific community. Admittedly there is much talk of atoms, molecules, ions, electrons, and other theoretical entities like orbitals but the actual and therefore decisive work is still done on substance samples.[18]

1.2 External Assessments – the Philosophical Stepchild

To get an impression of the current status of chemistry in philosophy, one can have a look at the relevant philosophy of science journals such as *Philosophy of Science*, *British Journal for the Philosophy of Science*, and *Zeitschrift für allgemeine Wissenschaftstheorie* as well as dictionaries. Even after what is now about three decades of flourishing professional academic philosophy of chemistry, the result of such a search for articles pertaining to chemistry is sobering. Apart from articles in the two specialized journals (*Hyle* and *Foundations of Chemistry*) and a larger series of omnibus volumes, the reader gets the impression that the traces of the philosophy of chemistry within the general philosophy of science have to date been relatively few and far between. This feels disproportionate particularly because the scientific output and societal relevance of chemical knowledge far outweigh those of all other areas of science. It would seem that "both" sides have tended

---

17 For example the 7th edition of *Römpps Chemie-Lexicon* contains the passage: "Chemistry is the science that concerns itself with causes and effects of the release, acceptance, or distribution of electrons between atoms or molecules and with the relationships between the energy levels of such electrons within the atoms or molecules." Römpp (1975, p. 548).

18 Note that Richter & Laitko (1962) also express such anti-cyclic preference ("macro-processes").

to keep to themselves, possibly due to the high degree of technical specialization (see below). Even in dictionaries, which are less dependent on current developments and are thus in a certain sense historically more stable, chemistry as a reference science continues to fare poorly. The renowned *Historisches Wörterbuch der Philosophie*, for example, devotes nine columns of text to physics and one to biology, whereas there is no such entry as "chemistry." Therefore the following question continues to be of fundamental interest to the philosopher of chemistry: What exactly is it that has, at least to date, made the study of substances appear so uninteresting or even irrelevant in the eyes of philosophical communities? Hopefully, addressing this question will also help to clear up the problem of what actually distinguishes chemistry and what its crucial characteristic traits are. The literature reveals several different approaches to answering the question posed here. The most important of these approaches are the immaturity thesis, the complexity thesis, the reducibility thesis, the dematerialization thesis, and the relevance thesis. In the present section I will discuss these theses in that order.[19]

The immaturity thesis assumes that a natural science becomes interesting for philosophy only in an advanced stage.[20] From this perspective chemistry as a modern science appears to be still too young for philosophical reflection (Janich 1980, 1992; Liegener & Del Re 1987). A comparison with other scientific fields is interesting. If we take a look at the biological and psychological disciplines, we may assume that the respective underlying sciences are not further developed or more mature than chemistry. And yet certain aspects that can be associated with them have been subjected to very intensive philosophical examination. Examples of these are the (historical) life-force discussion, the problem of evolution, the question of life in the strict sense, and the mind-body problem. One may also ask why physics questions, if we may call them such, were of interest to philosophers like René Descartes and Immanuel Kant long before the emergence of modern theories such as the theory of relativity and quantum theory. A more substance-related doctrine based on mutability and preparation would not have been excluded next to the more abstract mechanistic notions. Therefore scientific

19 The authors mentioned here call these theses. They are not necessarily proponents or even advocates of the theses I have summarized here for the purpose of discussion. Cf. Ruthenberg (1996a, 2024a).

20 Peter Janich distinguishes three graded suitabilities for disciplines to develop into sciences and states (1992, pp. 65–66; original author's emphases): "The discoveries of *theoretical suitability*, *empirical suitability*, and *philosophical suitability* of the natural sciences benefit physics in each case (and later biology as well) but disadvantage chemistry. Thus in whatever causal relationship to one another they may stand, chemistry initially lacks both academic reputation and academic institution."

maturity appears to me to play less of a role in the "philosophical suitability" of a scientific undertaking.

The complexity thesis maintains that chemistry's complexity and methodological diversity together with its confusing history render philosophical treatment difficult or even impossible. In other words, the experts in the philosophy of science and chemistry have not yet grasped the specific philosophical problems of chemistry (Ströker 1967, Plath 1990). Were complexity and diversity to prevent a fundamental philosophical reflection in the chemical realm, then that would pose the question of why it is that chemists themselves are well versed in their field but are hardly inclined to reflect on its fundamentals. Specialists in the sciences of substances are generally regarded as decidedly anti-philosophical. Chemistry has known only one single representative of its own guild who developed, in full scope, a philosophy of chemistry: Wilhelm Ostwald (see Chapter 4). At first glance it is hardly understandable why the complexity of a problem area should discourage an inclined philosopher from familiarizing herself with it. Other contexts are often not readily amenable to comprehension and yet attract attention, possibly for that very reason (how many hundred philosophical studies are there on quantum mechanics?). Is the philosophical status of chemistry thus so pitiful because not enough time has passed to find attractive theories, or are there, for reasons yet unknown, none to find? Those who primarily seek theories will have to conclude that chemistry actually exhibits a pluralistic diversity and that it therefore requires a different approach than, say, physics. To overstate the case, one can say that chemists do not search for theories in chemistry, rather they hunt for synthetic success and success in a wide variety of applications. To this extent, the complexity theory is indeed significant because it reveals that chemistry is different from mathematical physics. Without a change in perspective – away from static theoretical corpuscular thinking and toward dynamic substance-oriented thinking – one will not be able to do justice to the former.

The prejudice or, respectively, the conception that chemistry can at least theoretically be reduced to physics (theory reductionism) is widespread in broad areas of the natural sciences and unfortunately in the current philosophies of science as well. In the form of a reducibility thesis, the only aspect of chemistry that is of philosophical interest is that which is already being examined within the philosophy of physics (Primas 1985, Liegener & Del Re 1987, van Brakel 2014). The at times lengthy discussion as to whether chemistry is theoretically reducible to physics, a discussion that has accompanied the recent flourishing of the philosophy of chemistry from the outset, is one I prefer not to describe here.[21] This discussion has unfortunately fo-

---

21 Cf. here the collection of works by one of the protagonists of this discussion from the side of the

cused more on what chemistry is not and has placed disproportionate emphasis on the conventional physics-centric view of science. Thus too many publications ostensibly in the philosophy of chemistry have, by their own choice, placed themselves in the position of having to justify their existence. I expressly dedicate the present study to the question "What is chemistry?" in order to identify the metachemical particularities of this natural science. Supported by the main argument that even broad areas of physics itself are not reducible to any fundamental theory, I will limit myself for the time being to accepting the thesis that whereas chemistry should be theoretically compatible with physics, it is not reducible to the latter. If the reductionist approach were a sufficient answer to the question as to the reason for the deficient philosophical status of chemistry, then the consequence would be that it would be classified similarly to such fields as mechanics, acoustics, thermodynamics, the theory of electricity, and maybe fluid dynamics and meteorology: as philosophically irrelevant. There are certain areas within modern physics that imply epistemological and ideological questions for many philosophers, for example the theory of matter (particle theory), theory of relativity, and quantum theory and mechanics. In fact, quantum mechanics actually provides the conceptual framework for the modern theory of chemical bonds (see Chapter 8). However, there is almost no discussion of problems of quantum chemistry in the current philosophy of science. Reductionism is often encountered in one form or another, especially in physics. However, hardly anyone explicitly supports the opinion that its (assumed) reducibility makes chemistry philosophically irrelevant. A causal relationship between reducibility and philosophical unsuitability does not appear to exist as other philosophically interesting aspects are also subject to the assumption of reducibility with respect to other disciplines (such as biology and psychology). This thesis thus lacks the possibility of explaining the motivation for biological-philosophical and psychological-philosophical connections. Furthermore, it would not explain why chemistry outside of an assumed (current) reduction phase – disregarding for the moment existing exceptions – has been a philosophical stepchild.[22] The fact that experts from the science of substances occasionally apply reductionist arguments in their own explanatory conceptions and then immediately switch back to talking about macroscopic substance

---

philosophy of chemistry, Eric Scerri (2008), as well as his essay (Scerri 1996). At various points in this study I will come back to reductionism, if only implicitly.

22 Here it should be noted that the ancient natural philosophers and the alchemists made no distinction between (physical) matter and (chemical) substances. Cf. here especially Ströker (1967). My own attempts to elucidate the ancient "theories of chemistry" are Ruthenberg (1992, 1993, 2005) as well as Ruthenberg & Gerontas (2018). Thus if one can speak of a "stepchild problem" at all, then only in modern times, but in any case with respect to the 19th and 20th centuries. Cf. Ruthenberg (1994, 1995a).

characteristics certainly does not help the situation. In any case the reducibility thesis by itself does not sufficiently explain the philosophical undervaluation of chemistry.

Joachim Schummer advocates what I, again simplifying, call the *dematerialization thesis [Entstofflichungsthese]*, which he draws from his meticulous study of the history of philosophy, especially the history of natural philosophy and the philosophy of science. His interpretation is that during the course of the history of philosophy, which had its origins in the philosophical examination of substances, the material character of the world was later suppressed and formal aspects overemphasized. The science of the substances thus does not come to bear as a reference science in philosophy because its object is practically unknown there (Schummer 1996a). Schummer sees the beginning of dematerialization with Parmenides. In a certain sense Parmenides turns Heraclitus' experiential ideas conceptually upside down. Elisabeth Ströker, whose *Denkwege* continues to represent a standard work in early substance philosophy, expresses this aptly: "Whatever is not being, is not. The manifest diversity and manifoldness of existence thus sinks before Parmenides' thinking into appearance." (Ströker 1967, p. 22) Schummer describes the situation with a startlingly simple example, a sphere of water: The form philosopher will give (ontological) priority to the spherical shape, the substance philosopher to the liquid material, water. Whereas for the former the idea of the spherical shape is there first, the latter sees the substance with its specific characteristics (Schummer 1996a). Schummer advocates the view that the form approach has prevailed in the history of philosophy and that the substance efforts have been pushed "into an esoteric corner" to this day. In his view it is particularly the three old views about ignorance described by the presocratic philosopher Gorgias that play a role in answering the question of philosophical disinterest in chemistry: the ontological view (there are no substances), the epistemological view (we cannot detect them), and the view of the philosophy of language (if we could detect them, we could not speak about them). Added to this is the fourth attitude, prevalent in the current philosophy of science: "If there were a science that claims to speak about substances, then it would only be speaking about forms." (Schummer 1996a, p. 152)

This finally brings us to the relevance thesis. I understand this to be the epistemological alignment on the basis of certain ontological assumptions which I would like to describe in the following. In attempts at classification, the conventional order postulated for the "major" sciences is mathematics, physics, chemistry, biology, psychology, etc.[23] Entirely different meanings can be intended with such

---

23 Cf. for example Ostwald (1978) and Heisenberg (1989).

an order. Some authors (especially earlier ones) understand it as systematization of the objects of study; others see the focus on regularities and reliability of the statements made in the specific field; and for some it is a simple hierarchy of scientific character. Although such a classification does not comprehensively reflect all the essential connections between the sciences, one may conclude that mathematics is present in all of the subsequent disciplines yet neither physics nor chemistry nor any of the contents of the subsequent disciplines is present in mathematics. This fundamental asymmetry is also the reason why the reduction of chemistry to physics is normally discussed but not the reduction of physics to chemistry.

A classification system of disciplines as a framework in this regard becomes interesting particularly with respect to the various transitional areas as this is where the "major" philosophical topics lurk. I make use of the term "fulguration" in the sense that Konrad Lorenz (1903–1989) suggests. Lorenz rejects the term "emergence" as he feels the notion of "surfacing" from a fundamentally familiar and therefore predictable situation poorly suits his purposes because he seeks to describe the formation of that which is new and completely unexpected. Yet fulguration is nonetheless in essence an expression of emergence theory taken in a broad sense. Achim Stephan provides an initial description of the concept of "emergence" in his monograph on the subject:

> It often serves to characterize system properties that cannot be reduced in one way or another to properties of the system components or that are unpredictable or even exercise a "downward" causal influence (downward causation) on the system components. (Stephan 2016, p. XI)

On the basis of this description, it can be said that macroscopically observable substance characteristics arise from the components of the system as frequently unpredictable system properties. If these components are known, then the question soon arises as to how their interaction or structure relates to the system properties. Let us examine water as an example. Its empirical formula H_2O tells us that hydrogen and oxygen are present in a defined ratio. Under identical conditions, the two source substances have totally different characteristics. It is also entirely futile to want to explain even only one random substance characteristic of water from knowledge of these source substances. Melting and boiling points, behavior in solution, reactivity, etc. are and remain unexplained in this way. Although it may seem an awkward stop-gap, it is entirely legitimate to speak of emergence even in such very simple cases. The composition of a substance is certainly a necessary condition for its characteristics, but not a solely sufficient one (see Figure 2).

Fig. 2: Two metaphors for the interdiscourse concept of emergence: the unexpected "surfacing" of a phenomenon and the "fulguration" after Konrad Lorenz.

Konrad Lorenz applies his "fulguration" to a greater framework. He describes in a decidedly naturalistic manner the evolutionary steps of nature – for example in the origin of new species but also of those lying between the areas of reality or "layers of being" (inorganic, organic, mental, psychic).[24] The origin of the material world then corresponds to a fulguration resulting in the formation of the material substrate for everything else, the origin of life from non-life is a second fulguration, and the formation of sentient life forms and that of the mind are further fulgurations. If one then assigns the layers of being to the corresponding sciences, then almost without exception the order physics, biology, and psychology described above will result (mathematics as a formal or structural science drops out of this contemplation). Connected with the first fulguration and therefore with physics are the crucial questions of natural philosophy about the origin and structure of the world (the cosmos), i. e., cosmology and cosmogony. Problems of space-time (theory of relativity) and the structure of matter (radioactivity, ele-

24 See Lorenz (1977, especially Chapters II and III). He discusses and borrows the ontological four-layer doctrine of the German philosopher Nicolai Hartmann (1882–1950). In his *Der Aufbau der realen Welt* Hartmann (1964) claims the autonomy of these layers, which is similar to the position of a strong emergentism. – Lorenz was an exceptional natural scientist (Nobel Prize in Physiology 1973), yet also an avowed Nazi party member from 1938 on.

mentary particles, quantum mechanics) also fall within this realm. Yet where in this model (or similar models) does chemistry belong? Chemists apparently work in a field of science that seems scarcely relevant to the overall human orientation and worldview, despite the fact that the setting for their actions is undoubtedly the mesocosm. The next major topics, the epistemological *big points*, are the concepts of life and evolution which are "occupied" by biology, followed by the emergence of spirit and mind (or the mind-body problem). Localizing chemical phenomena in the mesocosm cannot be regarded as the reason for philosophical obliviousness to them. Now the model used here certainly cannot claim to have broad acceptance. I am using it here primarily to illustrate the essence of the relevance thesis. The fulguration or emergence metaphor may be applicable to smaller, localized phenomena in chemistry such as the formation of compounds with characteristics that cannot be predicted from knowledge of the constituents. Any way you look at it, the *major* topics, at least for the moment, are all occupied by the other sciences and therefore these sciences have had their "own" philosophy departments for a long time and are thus more attractive for analysis. In any case, in this way we arrive at an approach to explaining the philosophical undervaluation of the science of substances. This approach also provides a more plausible way to explain the widespread acceptance of the philosophical relevance of neighboring disciplines than do the other theses.

What could a strategy for the philosophical revaluation of chemistry look like? The simple postulate that a science would be relevant for philosophy when it (or better: its applications) has achieved societal (economic and industrial) significance seems to me to be insufficient. At least three other ways of improving chemistry's philosophical status are conceivable. One such strategy, admittedly a not particularly likable and also rather naive one, would be the "hostile takeover" of the life sciences. If it were possible to convincingly and comprehensively place the phenomenon of life on a chemical foundation, that is to eliminate the pertinent "fulguration," then biology would lose its philosophical fascination and chemistry would be enhanced accordingly. Efforts in such a direction have been present for some time and broad areas of modern laboratory biology are basically, at least in their methodology, now no different from chemistry although they bear the name molecular *biology*. Yet whether the capabilities of the sciences of substances will ever suffice to explain life entirely remains very doubtful. Konrad Lorenz addressed the subject of non-reducibility of living systems to inorganic matter, noting with skepticism:

> Yet the same applies in a like manner to the man-made machines that for that very reason provide a good illustration of the nature of the non-reducibility intended here. If one considers only their current physical framework of action, then they can be completely analyzed,

> down to the ideal proof of a successful analysis, down to complete feasibility of synthesis, namely practical producibility. Yet if one considers their historical, telenomic process of having become organs of homo sapiens, then, in attempting to explain their being this way and not another, one comes up against exactly the same non-rationalizable rest as with living systems. (Lorenz 1977, p. 54)

Secondly, a reverse and optimistic interpretation of Schummer's *dematerialization* thesis offers important perspectives: If it is possible to construct a substance-related philosophy then chemistry as the reference science may be expected to automatically become more the focus of discussion in the philosophy of science. In any case it is completely correct, actually even self-evident, with respect to the sciences of substances to give serious philosophical consideration to the significance of the concept of substance and all aspects related to it (see the discussion of self-descriptions above). Schummer himself contributed importantly to such a substance-related philosophy of chemistry and in many aspects I base my efforts on these contributions. The most monumental epistemological and ideological questions will perhaps be addressed to a lesser extent in the philosophy of chemistry, but certainly questions in the "middle range" of practical human experience, and in this sense the postulate of a scientific discipline with societal relevance should not be dismissed out of hand. These include such questions as what precisely *metabolism* is and what influence its natural and technical forms have on the environment. Roald Hoffmann, Nobel Prize winner in 1981, provides a clear message in this regard: "Chemistry is the truly anthropic science – our molecules can heal, and they can hurt, for they are on the scale of the molecules in our bodies." (Hoffmann 2007, p. 333) This obviously is a rather naive submicroscopic chemical comment – can molecules really hurt? – which leaves open whether the mesocosmic aspects mentioned are not best considered part of the life sciences – as long as the "fulguration" mentioned above has not been eliminated.

Third, a culturalist persuasion such as that advocated by Peter Janich (1942–2016) and his followers leads to the opportunity (indeed the necessity) for the operationalist reconstruction of chemistry as well as other areas. Culturalism follows the view that regularities are not discovered in or with nature but that they are constructed by researchers. This means that the practical background (the *practices*) of the sciences determines our concepts. It is obviously true that scientific results do not fall out of the blue. The philosophical reconstruction of practices is a valuable approach on the path to a philosophy of chemistry because its reference discipline is an experimental science par excellence. However, even culturalists must first specify what they want to understand as "chemistry." Aside from the purely reconstructive work, which would also come into question for carpentry or goldsmithing, where the actual philosophical topics lie must be determined in another way. A system that sees chemical concepts fundamentally based on the "Lebenswelt" and that refers only to purposes also risks ignoring the history and reality of chemical activities, for chemistry creates not only knowledge for dominion [Herrschaftswissen] but also knowledge for orientation [Orientierungswissen].

In this study I undertake an attempt to work out the distinctive features of the science of substances, in other words its metachemical fundamentals, which have been largely disregarded in the current philosophies of science. In so doing I attempt to examine and integrate historical and philosophical perspectives for the chosen aspects, including affinity, analysis, atomism and non-atomism, chemical species, chemical bond, chemical concepts, emergence, plurality, potentiality, reactivity, synthesis, temporality, and underdetermination. My fundamental position is empirical in the sense of perceiving and appreciating chemical practice. The conceptual point of departure and consistent point of reference is thus chemistry as a science of substances, expressly in contrast to its representation as a subdiscipline of microphysics. The essential historical reference authors for this study are "philosophizing" chemists such as František Wald, Wilhelm Ostwald, Friedrich Paneth, and Hans Primas. The unflattering yet occasionally not entirely baseless allegation that philosophers simply do not understand enough about chemistry to suitably evaluate it is one I choose not to dwell on here. On the contrary, I would like to counter it by presenting suitable case studies in sufficiently detailed form so as to stimulate discussion.

Fig. 3 Illustration of decomposition, isomerism, and mereological projection/fallacy using LEGO® blocks. If one assumes a ring-shaped structure (a "molecule"), then breaking it down completely will produce the individual constituents ("atoms"), bottom left. Yet even with precise knowledge of the characteristics of these constituents, one cannot reliably deduce the structure above and its other characteristics. The same set of blocks can also be used to construct the "isomer," bottom right. Compositionism as the mereological projection of parts on the whole is not the only methodological orientation in studying substances and can lead to fallacies.

2 The Berthelot Principle

Manufacturing, representing, realizing, fabricating, preparing, and producing are variations of manipulative interaction with the material environment, and they characterize the science of substances. In this chapter I will investigate the significance of synthesis in chemistry. The first section presents the manufacturing character of chemistry, which I will call the Berthelot principle. In the second section I will concern myself with the philosophy of the (chemical) experiment, and in the third section I will discuss historical examples.

2.1 Chemistry "Makes" Its Own Objects

The French organic chemist (and influential politician) Pierre Eugène Marcellin Berthelot (1827–1907) provided the appropriate credo, here in the original French:

> La chimie crée son objet. Cette faculté creatrice, semblable à celle de l'art lui-même, la distingue essentiellement des sciences naturelles et historique. (Berthelot 1876, p. 275)

I call this the "Berthelot principle."[1] This designation is justified because it was Berthelot who first presented this specific metachemical statement in a clear and programmatic form although earlier chemists are known to have made similar pronouncements. At first this may sound like the preparative chemist exaggerating the virtues of his own field of work. Yet Berthelot by no means limited his studies exclusively and one-sidedly to one single area of the sciences of substances. He worked very intensively and influentially in the history of chemistry, especially with respect to alchemy, and in these studies he emphasized the manufacturing aspect that can be found in the precursors to industrial chemistry such as metallurgy, pottery, dyeing, painting, food production, medicine, ceramics, and glass making (Berthelot 1909, pp. 90–101). The Berthelot principle is thus this scholar's crucial premise, and its significance for the philosophy of chemistry can hardly be

1 Bernadette Bensaude-Vincent and Jonathan Simon (2008, p. 99) give Chapter 6 of their book *Chemistry – The Impure Science*, which has some similar content to the present study and also deals essentially with Berthelot, the title *Chemistry Creates Its Object*. The authors provide the following translation:"Chemistry creates its object. This creative faculty, akin to that of art, forms an essential distinction between chemistry and the other natural or historical sciences."

https://doi.org/10.1515/9783111401041-005

underestimated.[2] In the introduction to *La Synthèse Chimique* the author criticizes what he feels is Antoine Lavoisier's (1743–1794) one-sided attitude as follows:

> As Lavoisier nearly a century ago based chemistry on the study of the continual separations of which the bodies are capable, and made the final product of these separations, i. e., the simple bodies, the foundation of the science, he arrived at the definition of chemistry as the science of analysis.
>
> This definition, however, is incomplete; it only expresses one side of the problem. For by way of the repeated separations we have penetrated the essence of the bodies, and we are now presented with the additional task of rejoining what was separated and restoring what was destroyed. (Berthelot 1877, pp. 1–2)

It is an indisputable fact that Lavoisier played a pivotal role in establishing the modern concept of the element. In this context a chemical element is a substance that cannot be further separated into simpler substances. However, this logical and formal notion of analysis can rarely be achieved in laboratory practice. The "final products of separations" were neither produced and counted up in the laboratory at that time, nor are they now (see Chapter 5). Lavoisier in particular should have known better, seeing as he was the one who was primarily responsible for the oxygen theory of combustion, which is still valid today. When a substance burns in air, it reacts with oxygen and produces combustion products we refer to as oxides. Hardly anyone would refer to the formation of these combustion products as separation even though the original substance disappears, i. e., is "destroyed." It is only by conceptual analysis that we can deduce the influence of oxygen from weighing the products created.[3] Berthelot means something else when he makes the following statement, but this can also apply to other forms of synthetic activity in support of chemical examinations:

> It is only this potential of the synthetic process that gives chemistry its true character; it distinguishes it from the other natural science disciplines that are based solely on dissection and ensures a higher degree of reliability for its results. (Berthelot 1877, p. 2)

This "higher degree of reliability" is nothing more than the comparison with reference standards that is possible only after an appropriate synthesis. Division alone without the opportunity for comparison would be preposterous. By "divi-

---

2 Mareike Frevert provides an informative representation of the significance of Berthelot to the philosophy of chemistry (Frevert 2014, pp. 123–132). In her study, the author sympathizes with the "surrationalist structuralism" of Gaston Bachelard.

3 Today, this fact is still used in elementary analysis for the routine examination of compositions (cf. the discussion in Chapter 5).

sion" Berthelot does not yet mean the breaking down of substances but rather the dissection of objects of study. He gives examples of this, which he systematically elucidates.

Let us examine one of these examples, the examination of a lemon. The lemon is an aggregate of several different types of bodies. When one presses it out, one is left with the sour yet slightly sweet juice and the "pleasant smelling" peel. If the liquid portion is further examined in the same manner that the components are separated but not altered (here he does not say how), "then we obtain a certain number of more immediate or more specific components such as citric acid, which gives it its sour taste, dextrose and sucrose, the cause of the sweet taste, an albuminous substance, salts, etc., and finally water, which contains the substances mentioned dissolved within it" (Berthelot 1877, p. 4). All the "isolated bodies" each have unchanging and specific characteristics; one "[…] cannot separate them into several new substances without destroying all of their characteristics." If one now examines the peel, one will analogously (here, too, the author does not reveal exactly how) obtain a volatile oil that gives the fruit its aroma, a yellow body soluble in "ether" (diethyl ether) that represents the fruit's yellow color, and a wood-like material that represents the bulk of the lemon peel.

Berthelot states that the combination of all these substances would constitute the original fruit (that is in the sense of the whole that consists of its parts) and that the first part of the analysis would thus be completed (Berthelot 1877, p. 5).

> Once the analysis has reached this first stage, it devotes its further activity to the immediate components themselves, penetrates deeper into their composition, and finally reduces them to a certain number of indivisible elements. (Berthelot 1877, p. 6)

Thus, in the course of the second analytical task, the qualitative elementary composition of the individual constituents is examined. The citric acid for example "contains" the elements carbon, hydrogen, and oxygen; the albuminous substance carbon, hydrogen, oxygen, and nitrogen.

> The second analysis […] uses the strongest agents, and processes and changes the immediate components themselves; it proceeds systematically to their complete destruction until the elements, i. e., the components incapable of further separation, have been isolated. (Berthelot 1877, p. 7)

Various questions and comments arise when regarding this instructive and aesthetic example. First, the inclined specialist would ask: What methods are used to separate the constituents of the individual compartments of the fruit? Presumably, one of the old basic operations, distillation, has been used here at least to some extent. Yet how the solid residues could be then separated remains unanswered. The method Berthelot has thought up is quite probably significantly more complex than would initially appear (yet maybe he underestimates it as well). Second, where the knowledge of the individual components named here comes from remains completely unclear. Maybe one learns in school that lemons contain the acid named after citrus fruits, but which of the white powders that remain after evaporating the water is it? How can we know about the characteristics of citric acid? Third, why the analysis has to proceed to the second stage at all is questionable when we may already have long since acquired a lot of information, maybe the most important information, about substances that we require. Fourth, an elementary analysis, both then and now, is not really destruction, rather as a rule – for example in electrolysis – new substances are created which are then examined in greater detail. One can only speak here of "isolated elements" in a very abstract conceptual sense.[4] In reality they do not exist in the experimental setting. Fifth, one final question in relation to this aspect: Is the information about the type and quantity of the smallest building blocks really so fundamental that one must feel compelled to regard it as a necessary condition for synthesis? Berthelot initially appears to give priority to analysis:

> However, if it is therefore correct that analysis forms the starting point of chemistry, then it [analysis] cannot simultaneously be its goal and its purpose [...]. A complete chemical knowledge of nature can only be achieved by combining the opposing methods of analysis and synthesis. (Berthelot 1877, p. 2)

As we have seen, the characterization of substances begins with observing an object more closely, selecting which parts of this object one wants to examine, and then actually "taking it apart" to identify, if at all possible, the individual constituents present after this separation. This separation is what is referred to in modern analytics as sampling (including sample preparation; cf. Danzer 1992). When one begins with a complex examination object, one may call this method "analytical"; yet because its content is closely related to the manufacturing processes, I would regard it as "preparation" and thus more a synthetic aspect.

The question posed above under "fifth" contains a general problem of compositionism: Is it sufficient to know the precise composition of a substance in order to characterize it and distinguish it from other substances? Here, mere reference to the isomers in organic chemistry, in which elementary analysis leads to the same empirical formula, convincingly answers this question in the negative (cf. Figure 3). It is similar to tearing down a building: Even with the most painstaking and me-

4 Claus Jacob (2001) discusses the very important distinction between linguistic (conceptual) and experimental (operational) forms of analysis and synthesis.

ticulous "dismantling," while it leads to countable, measurable results (such as a certain quantity of shingles, bricks, beams, boards, mortar, windows, doors, etc.), it is only the "resynthesis" of the complete structure that creates an impression of what has actually been examined here.[5] Whether or not we prefer to classify the sample preparation and the many "dismantling" procedures under the expression "analysis," the results of the experiment, both qualitative and quantitative, are confirmed by comparison with experimentally established (manufactured) references. Berthelot also confirms this for both inorganic and organic chemistry.

This critical discussion is by no means intended to belittle Berthelot's fundamental argumentation and the Berthelot principle in particular. On the contrary, he is completely correct in saying that by restricting ourselves to analysis as he depicts it we are unable to comprehensively describe the chemical substances. There is yet another point that is important above and beyond this (however, one which Berthelot does not discuss): Continually breaking it down increasingly moves the experimental substrate farther away from the realm of substance so that even the synthetic confirmation processes become decreasingly useful. When one finally arrives at the elementary particles, one has passed beyond the chemical range of influence and detection.

Berthelot developed his theses in the philosophy of science based on his professional experience as an organic synthesist. The creation of organic compounds experienced an enormous surge in the 19th century (cf. Schorlemmer 1889) and he himself greatly influenced this with his systematically developed and expedited "total synthesis," that is the manufacture of desired substances from simpler raw materials. The French philosopher of chemistry Gaston Bachelard commented on the situation in *La Philosophie du Non* in the year 1940 as follows:

> In the initial period of organic chemistry one liked to think that synthesis only served to verify the accuracy of analysis. Now it is rather the opposite. Every chemical substance is only really identified at the moment of its reconstruction. (Bachelard 1940, p. 70)

Thus when Berthelot says that perfect chemical knowledge of nature can only be achieved by "uniting the opposing methods" of analysis and synthesis, that does not mean we can deduce any symmetrical equivalence of these concepts and their correlates. The Berthelot principle is an important condition for chemical work, both in synthesis and in analysis (see Chapter 5). I will permit myself to

---

5 This image is rather skewed because a direct comparison between the two buildings is not possible. This is also related to the fact that a building is a thing, and while substances in a certain sense can exhibit individuality (see Wald 1899), they do not have to have the nature of a thing (cf. Soentgen 1997, Schummer 2017).

add a heretical "sixth" to the discussion of the lemon examination: What would such an analysis look like for a fruit that is completely unknown to the laboratory staff?

2.2 Chemical Experimentation – Extending Reality

Classical philosophy of science usually describes experiments as means of proving or disproving hypotheses. Yet the theory of experiments can and should be considered more broadly in the case of chemistry. The substance perspective requires us to critically examine the fact of the emergence of that which is new, and not only in conceptual and theoretical form. A number of authors[6] have lamented the underdeveloped state of the philosophy of the experiment, which is quickly confirmed by a look at overviews, introductions, and dictionaries in the philosophy of science. Usually the treatment is limited to discussing or mentioning the famous *experimentum crucis.*

Michael Heidelberger (1997, 1998, 2003) offers an interesting approach. He devises for experiments and instruments a group of three purposes that may be considered as features in addition to (or together with) the conventional verification of a theory:

1. the purpose of expanding experience (*productive function*),
2. the purpose of structuring phenomena (*representational or structuring function*), and
3. the purpose of controlling conditions (*constructive or imitative function*). (Heidelberger 1997, p. 5; original author's emphasis)

Heidelberger sees the productive function realized in equipment such as the air pump or the electrostatic generator. It is also interesting for our discussion that the author lists here the particle accelerators that "produce new building blocks of matter." Another type of productive function is found in telescopes and microscopes, which do not produce phenomena but by their character improve sensory perception. In addition to this, Heidelberger also names those instruments that separate known phenomena into components in some manner as prisms and spectroscopic apparatuses do. Representational instruments such as clocks, scales, and thermometers have an ordering function in this system in that they produce connections such as the artful transformation of the sensation of heat into a calibrated

6 Cf., for example, Hacking (1982) and Heidelberger (1998).

expansion of mercury.[7] Finally, Heidelberger finds that experiments are imitative when a phenomenon is rendered controllable in a desired manner. Examples include the creation of lightning to test lightning rods or the simulation of winds in a wind tunnel. Heidelberger draws the following conclusion from his study:

> For all three aspects of experiments, the productive, representational, and constructive, there is […] a certain autonomy with respect to theory. Experimenting is only possible "against a background formed by expectation" yet with purely productive or constructive experimenting it does not require any theoretical interpretation. This shows […] that experimenting also possesses a genuinely creative function that is independent of theory. The experiment, in addition to verifying theories, also serves to create reality. (Heidelberger 1997, p. 9)[8]

A survey of the relevant literature reveals that hardly any of the philosophers of science who have investigated the significance of the experiment in greater detail have thought to include chemical empiricism. The manufacture and preparation of substances are airbrushed over.[9] Using Heidelberger's system, however, it would be easy to present the synthesis of chemical substances as nothing less than a prototype of productive experimenting. Is the manufacture of substances something scholars can ignore? Why is the production of "new building blocks of matter" in particle accelerators worthy of mention but not, just to name one example, the systematic and artful manufacture of life-saving agents? The extension of reality (the title of Heidelberger's article) is a crucial characteristic of chemical activity and can therefore be rightfully regarded as a leitmotif for the discussion of the Berthelot principle. Future theories of scientific experiments will in any case no longer be able to ignore the manufacturing character of chemistry.

Chemistry "makes" its epistemic objects itself. I apply this fundamental metachemical concept to the entire spectrum of the science of substances. I will now proceed to discuss the views of one of the few modern philosophers who has ad-

---

7 A book title like *Inventing Temperature* is in this sense particularly appropriate, see Chang (2004).

8 Heidelberger's taxonomy has been reviewed in the history and philosophy of the sciences; some responses have been largely critical (Hentschel 2000, p. 25), some assenting (van Fraassen 2008, pp. 94–96).

9 Just to mention one example: In the collection edited by Lorenz Krüger with the attractive title *Die Entstehung des Neuen* [The Emergence of the New] Thomas Kuhn, for whom the so-called "chemical revolution" was known to be the anchor of his interpretation of the history of science, examines exclusively analytical and theoretical aspects of chemistry, yet by no means chemical synthesis (Kuhn 1978). Hugo Dingler (1928), too, does not represent an exception in this respect in his work on the philosophy of experiment. Cf. also Ursula Klein's critical attitude to Ian Hacking's theory of the experiment in Klein (2008, p. 42).

dressed this crucial topic for chemistry seriously and at length.[10] I will begin with an essential result of the study *Die Rolle des Experiments in der Chemie* by Joachim Schummer, which he presented at the first Erlenmeyer Colloquium in Marburg, Germany. He refers to the specifically chemical perspective and states:

> In the attempt to more sharply define this *substance perspective,* it takes on methodological characteristics lacking in traditional models of knowledge: action aspects of actively constituting the object, actively creating the context, and performing distinguishable processes. With the systematization and clarification of the applicable rules of action, the substance perspective becomes a scientific examination method; it becomes the *experimental* method of chemistry. (Schummer 1994, p. 48; original author's emphasis)

Here the purely passive aspect of the earlier natural sciences in which one attempted to "read in the book of nature" is attacked, and even the measuring activities of experimental scientists can only be a part of "actively constituting the object, actively creating the context, and performing distinguishable processes." The first point in particular, actively constituting the object, manifestly aligns closely with the Berthelot principle as I have characterized it here. The idea of manufacturing epistemic objects has its precursors in the modern study of science. Two staff members of what was then the Max Planck Institute for the Study of the Scientific-Technical World in Starnberg, Germany, describe the "structures of pre-paradigmatic science" with the following words in a major study of the origins of autonomous scientific development:

> The object of experience and its possible variations are [...] "manufactured" by the process of experience itself. The phenomenon becomes the effect. Thus the difference between natural things and technical artifacts disappears for natural science. (Böhme & van den Daele 1977, p. 188)

This description applies unconditionally and particularly well to chemistry, and even to its "pre-paradigmatic" stages. Yet the manufacture of a science's own epistemic objects and thus, among other things, an extension of reality apparently lies beyond the chosen perspective, as is also the case with Heidelberger and the majority of experiment philosophers. Nonetheless, the same authors refer in another study more specifically to experiments from chemistry and differentiate "analytical" and "synthetic" devices and processes (Böhme et al. 1978, pp. 354–355). Measuring devices such as a clock, thermometer, calorimeter, barometer, volt and amperemeter, and processes such as Liebig's elementary analysis and dextrose

10 Historians of chemistry are familiar with what I refer to here as the *Berthelot principle.* However, the recent philosophy of chemistry has been more concerned with topics of microphysics.

detection using Fehling's solution are termed "analytical," whereas devices such as the vacuum pump, electrostatic generator, voltaic pile, electromagnet, and processes like the synthetic manufacture of dyes are "synthetic." The authors continue:

> In general any experimental configuration that serves to represent an effect or a substance, insofar as it is standardized, can be regarded as such a synthetic device or process. (Böhme et al. 1978, p. 355)

Although it would have been possible to make the connection to the philosophy of chemistry – the authors discuss Gaston Bachelard for example – this approach does not go beyond simply "representing" a substance.

In stark contrast to this, Joachim Schummer devotes a large part of his study *Die Rolle des Experiments in der Chemie* to the substance perspective and its pivotal role:

> My thesis is that that which has substance can be grasped conceptually only from a certain intentional approach and that this approach in the methodological clarification of chemistry can rightfully be termed an experimental method. (Schummer 1994, p. 36)

Accordingly, it is not at all obvious what substances are, rather this is only deduced by actively dealing with objects of study (such as Berthelot's lemons). Sample selection, sampling, sample preparation, and also sample preservation and fixation become particularly important. Other conditions are stipulated, which also form the basis of the conventional empiricist model of knowledge. These include repeatability and the exclusion of disturbances and illusions. Yet for chemistry the requirements are even stricter:

> An experimental science is now faced with the particular task of explicating the *contextual* conditions and systematically and extensively identifying their influence on the experiments. Only then can we speak of an experimental study in the strict sense. (Schummer 1994, p. 36; original author's emphasis)

The concept of substance is thus significantly less abstract than the concept of matter. Schummer makes a typological suggestion (Schummer 1994, p. 39) for the contexts of the experimental substance perspective he mentions. He lists the following contexts (examples in parentheses): 1. *mechanical* (Mohs scale hardness, fracture strength); 2. *thermal* and *thermodynamic* (melting point, effective heat capacity); 3. *electrical, magnetic,* and *optical* (conductivity, magnetic susceptibility, refraction index); 4. *biological* and, respectively, *ecological* (toxicity, ozone-depleting potential); 5. *chemical* (ability to oxidize with oxygen, water solubility).

Schummer notes that one could combine contexts 1–3 as physical substance characteristics, which is in line with customary usage. In the discourse of the philosophy of science this has occasionally led to the view that chemistry is based on physics. However, one should not overlook the fact that the interplay of all the contexts is what is important for the comprehensive description of substance.

Schummer makes an important comment with respect to the correlate of a classification of a characteristic. Melting and boiling points, for example, are listed in tables as "physical constants." We speak of these specific characteristics in such a manner as if the listed substances had these characteristics in themselves in the way that something is red or large or old. However, all of the substance characteristics of the contexts described reveal themselves only at the moment the experiment is performed. Substance characteristics are therefore disposition concepts (or predicates). In the cases of contexts 4 and 5, the situation is even more complicated. In these cases, the questions arise as to which of the involved components one assigns reactivity to and which the context role (this aspect becomes particularly thorny when examining catalysis). Schummer feels that we must speak of *disposition relations* here. The assignment of characteristics to individual substances is obviously a simplification attributable to common language usage.

A brief look at the history of natural philosophy reveals that substances have been treated as individual objects of our environment since ancient times:

> All these bodies differ from each other, firstly, in the particular ways in which they can act on the senses (for a thing is white, fragrant, resonant, sweet, hot or cold in virtue of the way it acts on sensation), and, secondly, in other more intrinsic qualities commonly classed as passive – I mean solubility, solidification, flexibility and the like, all of which, like moist and dry, are passive qualities. (Aristotle 1952, p. 339 [385a])

In the fourth book of *Meteorology* Aristotle describes a few disposition predicates, referred to here as "passive behavior."[11] The characteristics that he regards as "active" are directly perceivable with the senses (and are hardly touched on by Schummer).

On its path of experimental substance study, chemistry increasingly switched over to developing a hierarchy of substances during the course of the 18th century.

> It is not by chance that particularly the active aspects of the experiment, those geared to changes, hold entirely new potential that has remained completely inaccessible to epistemology and the philosophy of science. (Schummer 1994, p. 40)

11 Cf. Böhme (1980).

Especially the operation of *breaking down* is geared to change. One encounters this process prominently in Lavoisier's "analytic" definition of element. I have already mentioned it in the discussion of Berthelot's methodology. By means of this process one can verify, at least in principle, whether the substance at hand is a pure elementary substance in the modern sense. Schummer in his fundamental work *Realismus und Chemie* provides a more precise definition of breaking down:

> An operation on a substance sample is referred to as breaking down a substance when it results in at least two substance samples n_i, n_j which are distinguishable from each other and from the initial substance sample from the substance perspective. (Schummer 1996, p. 170)

We can regard the electrolytic separation of water (n) into its constituents oxygen (n_i) and hydrogen (n_j) as an example of breaking down a substance. It is not necessary to fully identify these reaction products in order to differentiate them from the initial substance. This is an interesting and important borderline case of chemical analysis as the goal is only to verify whether or not the substance sample is a composite. Yet with respect to the products, this is a synthesis (see Chapter 5).

In addition to the individualization and classification of individual substances, the next logical step was the creation of order. Chemistry took a great leap forward with the differentiation of pure and mixed substances as well as elementary and compound substances now presented in every introductory chemistry lecture. Schummer refers to the content of the Berthelot principle (here without naming Berthelot) in his discussion of the aforementioned manufacturing character of the chemical experiment. His following statement can be regarded as a modern definition of this principle:

> Modern chemistry stands as nothing less than the paradigm of a science that synthesizes its own objects and very rapidly expands its object range. (Schummer 1994, p. 43)

The synthesists responsible for this manufacturing character do not do this solely because of industrial and technical interests or comparable purposes, but because they, as Schummer writes, "feel committed to the academic natural science of chemistry." Thus to some extent they act from an epistemological perspective. The manufacturing character is grounded in the logical structure of the chemical experiment:

> Note that this manufacturing character is not an epiphenomenon but a necessary condition for us to be able to speak of chemical experiments at all. (Schummer 1994, p. 43)

Schummer draws an important conclusion from the manufacturing character of chemistry and, as he calls it, the *web of relations* developing out of it:

> Preparative chemists are to a certain extent realists by profession with respect to the substance entities they design theoretically. For with the theoretical design of a substance entity they simultaneously demonstrate its possibilities for realization in the sense of a guide for experimental synthesis that they can immediately put into practice. (Schummer 1994, p. 48)

The Berthelot principle (the manufacturing character) is thus, according to Schummer, the basis for the "non-naive experimentalist realism" in chemistry.

2.3 Examples

Three examples are intended to illustrate the metachemical manufacturing character of the science of substances: The early manufacture of acids by "distillation," Wöhler's urea, and the Gomberg radical.

(1) The historian Bruce Moran describes chemistry very aptly:

> Process in action. It makes things happen. Sometimes it separates and considers boundaries. Sometimes it combines and queries about connections. It leads to control, to the artificial construction of useful objects and to claims of power. And all these things – processes, practices, as well as theories – are important to the pursuit of natural knowledge. They are all parts of the diversity of learning that helped to create what we call the Scientific Revolution and that affect the doing of science still. (Moran 2005, p. 188)

In his book *Distilling Knowledge* Moran examines the late alchemy of the 16th and 17th centuries in particular and its influence on the so-called "chemical revolution," that is the transition to compositional modern chemistry in the 18th century. "Distilling knowledge" is a particularly appropriate title as it makes clear that distillation (which already was indirectly mentioned in Berthelot's lemon examination) is, first, a genuinely chemical operation and beyond that can be regarded as belonging to the synthetic processes, especially when chemical reactions in the strict sense are in play.

Since humans have practiced the controlled use of fire, they have used it not only to cook food but also to transform other materials into artificially altered products used for various purposes. One example of these transformations is the manufacture of wood tars from birch or the wood of coniferous trees. Distillation was developed from cooking food in a watery medium, hardly as an intentional invention; a substance substrate that rose as the vessel was heated was found to reappear (in modern terms: condense) in another place when cooled, perhaps in an altered form.

All possible materials, primarily those of vegetable origin, were subjected to such a procedure, which suggests that this is why chemical species such as ethanol and acetic acid became known very early in nearly every culture. Acetic acid, now thought to have developed from the fermentation of wine, is accordingly the first acid in history. Naturally, one should be particularly careful with using modern substance designations and the theoretical concepts they entail. Since Plato at the latest, all liquid substances (in Europe) have been classified under the element of water. Taking appropriate care with terminology as mentioned above, one can nonetheless assume a certain degree of individualization of certain chemical entities, frequently even with proper names, even though chemistry as a discipline or an explicit concept of substance did not yet exist.

Distillation is strictly speaking a substance transformation, for a substance in the liquid state is organized or structured differently than in the gaseous state. This may seem like splitting hairs and is in any case unusual, but it follows logically when we place the focus of chemical interest on the concept of substance. Only the composition of the gas and the liquid are the same; however, they must be verified by further operations. Wilhelm Ostwald proposed the concept of *hylotropism* for discussing this and similar problems (Ostwald 1907, pp. 259–262; cf. Chapter 4). He describes substances as hylotropic when they do not change their composition when changing their phase (i. e., melting, evaporating, sublimating). Even pure substances can change certain characteristics. Hylotropic substances do not necessarily have to be pure substances, for there are liquid and solid mixtures that exhibit hylotropic behavior.

When such a transformation of pure substances occurs only by operational intervention, then the situation known traditionally and in alchemy as "distillation" is in any case something synthetic. To explain what I mean by this, I will now briefly examine another historical example from the fascinating realm of the acids. As was mentioned, vinegar (acetic acid) was historically the first known acid and with its behavior (its characteristics) provided the prototype for the entire substance class including phosphoric acid.

Phosphoric acid is a latecomer in the group known as the "mineral acids." Researchers learned of its existence only around 1700 although we can assume that it had occurred unnoticed in laboratories and workshops at one time or another. The manufacturing and preparation techniques of the acids were characterized from the outset by distillation (occasionally "dry" distillation). Solid raw materials (minerals) were intensely heated with water or alcohol and the breakdown products were cooled, condensed, and collected. Here is a classic recipe:

> Saturate a pulverized skull with betony water or spirits of wine. Place that in a pot. Pour peony water into the lower pot, start a fire, and distill. (Libavius 1597, p. 424)

This recipe for distillation from a water bath will hardly have produced phosphoric acid but maybe a few of its esters. With a little imagination one can well suppose that pulverized bone was also subjected to intense heat. The formation of phosphoric acid – long before the modern element of phosphorus first became known – was thus quite probable. The German term "Knochensäure" ("bone acid"), used by a few authors as late as the 18th century (and listed in Grimm's dictionary of the German language), thus seems plausible and logical. In his famous *Traité élémentaire de chimie*, Lavoisier describes the production of phosphoric acid, which had since become a standardized procedure, as follows:

> The best method of obtaining this acid in a state of purity is by burning well purified phosphorus under bell-glasses, moistened on the inside with distilled water; during combustion it absorbs twice and a half its weight of oxygen [...]. This acid may be obtained concrete, in form of white flakes, which greedily attract the moisture of the air, by burning phosphorus in a dry glass over mercury. (Lavoisier 1789, p. 226)

As with many other classic acids, the discovery and manufacturing history of phosphoric acid can be traced back to the use of organic (biological) material. Phosphorous, which when burned yields a precursor of the acid, was first obtained from urine and later became an element in the modern sense. Nitric acid also belongs to this transitional area between chemistry and biology. It had been known many centuries longer and was obtained similarly by intensely heating saltpeter (from human and animal excrement) with alcohol or water.[12]

Unlike carbonic acid, phosphoric acid is not a chemical "spirit," meaning it exists in pure form. At room temperature it is a solid that is both hygroscopic and soluble in water in large quantities. However, this substance does not occur naturally in pure form but must be artfully prepared. In spite of this it is not artificial in the strict sense: Marcelin Berthelot was entirely correct when he stated that chemistry makes its own research objects, and it does so well beyond organic chemistry. Phosphoric acid is a "medium-strength" acid that has four bonding points per molecule and is therefore predestined to form "esters." Yet its acidic character is not the most important feature of this chemical substance. Rather it is this variety together with the usually very high chemical stability of its derivatives that defines its particular capabilities. Here we think of our bones, the nucleic acids, or also, unfortunately, the chemical warfare agents sarin and tabun.

It should be noted that the form of "distillation" described has for many centuries been a synthetic, that is genuinely *chemical*, means of manufacturing *and* studying substances. (It was only since the advent of precise temperature measure-

12 Cf. Ertl & Soentgen (2015, Chapter 2) and Ruthenberg (2020, pp. 61–82).

ment in the chemical laboratory and in process engineering in about the mid-19th century that the technical jargon came to refer to distillation as a "physical" separation technique).[13] Bruce Moran describes the significance of the historical, that is "synthetic," distillation very elegantly when he states: "There is something about doing distillation that combines action and reflection in such a way as to produce a feeling of unity and knowing." (Moran 2005, p. 12)

(2) The second historical example can be very precisely dated and has developed an enormous effect that goes far beyond the simple manufacturing aspect. This is the famous urea synthesis that Friedrich Wöhler (1800 – 1882) published. This study produced the

> unexpected result that when cyanic acid is combined with ammonia, urea is formed, a fact that is also notable insofar as it represents an example of the artificial creation of an organic, indeed so-called animal, substance from inorganic substances. (Wöhler 1828, p. 253)

The chemist (and well-known historian of chemistry) Paul Walden (1863 – 1957) provides a very lively and fitting discussion of Wöhler's results and their consequences about a century later.[14] The manufacture of a substance that until then had been assigned to the organic realm, that is to living beings,[15] from inorganic raw materials was nothing short of a sensation, and Wöhler himself could hardly believe his result. In his article with the subtitle *Ein Jahrhundert der chemischen Synthese* Walden discusses in depth the connection between Wöhler's discovery and the vital force (*vis vitalis*) that had been propagated by Jöns Jakob Berzelius (1779 – 1848), one of Wöhler's teachers. Walden mentions such thinkers as Berzelius and Schopenhauer and summarizes that "exact research" and "strict philosophy" concurred in assuming the existence of a vital force. Among other things he also describes how Wöhler in his correspondence with his Swedish teacher had

13 As regards the history of the development of temperature measurement in which substance characteristics also play an important role, cf. Chang (2004).

14 The Latvian Walden, a student of Wilhelm Ostwald, was also active in synthesis; a form of organic substitution is still referred to today as the Walden inversion. Additionally, he was also interested in systematics (see the discussion of radicals in the following example and the Ostwald-Walden rule for acids and bases), cf. Walden (1924) and (1929). Walden was a Nazi party member from 1934 on and occasionally speculated and held forth even before that about a "German manner of chemical research," see Maier (2015, p. 330). His achievements in the history of chemistry are nonetheless significant.

15 Urea is one end product of nitrogen metabolism in many higher life forms. Healthy human kidneys excrete 30 g daily of what is now known as the diamide of carbonic acid with the empirical formula $CO(NH_2)_2$. It is the raw material for creating saltpeter by microbial oxidation.

been very reserved with respect to the vitalism issue. Walden particularly emphasizes one aspect (a "dramatic aspect") I feel is typical of chemistry:

> *[N]ot* the *theory, not* the scientific *search for this synthesis,* but teasing *chance* led to the result: *an unsuccessful experiment,* insofar as the salt predicted by the theory and intended by Wöhler (ammonium cyanate $CNO \cdot NH_4$) was *not* produced, but instead a rearranged product of it, urea $CO(NH_2)_2$! (Walden 1928, p. 836; original author's emphasis)

Essentially an inorganic chemist, Friedrich Wöhler, first, had not foreseen the consequential result of his experiment and, second, had himself in no way proposed this result as an argument against vitalism. This restraint is primarily attributable to the fact that the raw materials for his experiment originated in the biogenetic realm, something he was aware of. Yet it also reflected appropriate respect for his teacher Berzelius.

Wöhler's discovery went down in the history of chemistry as the beginning of the demise of chemical vitalism, and many authors see it as the actual starting point of synthetic organic chemistry. However, the occurrence of living material has not been "explained away" by Wöhler's discovery.[16] Although both carbon chemistry and the allied fields of biochemistry and molecular biology have since made great individual advances, a "chemical explanation" of life remains elusive although ambitious theses are repeatedly proposed, claiming that the compositionist examination of living material would be an important step in the right direction. Although one must admit that a decompositionist examination would represent a necessary requirement for deciphering the miracle of life, it is certainly *not* sufficient. In our philosophical embarrassment we can attempt to describe the situation using concepts such as emergence, but in the final analysis nothing changes: We cannot explain the phenomenon of life in chemical terms. Vitalism has thus essentially not been refuted.

(3) From the perspective of modern chemistry, radicals are substances whose molecules have unpaired electrons. The history of the radicals can be divided into three periods which I refer to as the speculative, synthetic, and electronic periods. The example described here marks the beginning of the second period.[17] The term radical ("root") was first used in the realm of chemistry by Antoine Lavoisier (1743–1794), who used it to refer to the postulated essential constituents of acids. Jöns Jacob Berzelius (1779–1848) used the term as an "element imitator," that is to say as a speculative ordering principle. The program of research into

16 Cf. the detailed discussion in Weyer (2018b).

17 I have discussed this synthetic case study in greater detail in Ruthenberg (2015).

the manufacture of radical substances began with the work of Moses Gomberg (1866 – 1947), and since the 1930s discussions in the field, set in motion by Gilbert Lewis about a decade earlier, have centered around the "electronic characteristics" of chemical entities, although competition to create new (and also radical) substances continues to this day.

As a young post-doctoral researcher in Victor Meyer's laboratory in Heidelberg, Gomberg successfully synthesized tetraphenylmethane, the "perphenylated" methane $C(C_6H_5)_4$, something Meyer had been previously unable to do. Gomberg's successful approach was to split triphenylmethylazobenzene (Gomberg 1897). In so doing he had essentially already made a name for himself in the guild of preparative chemists. Yet he went on to greater achievements, and ironically this happened because of a failure. Back in Michigan, he pursued the goal of producing so-called "arylalkanes." These are compounds of simple hydrocarbon chains with ring-shaped "aromates." He thought it expedient to attempt the synthesis of fully phenylated hexaphenylethane in order to verify the test of the ability of carbon atoms to form four bonds when fully nitrated.[18] Following a similar principle practiced earlier with tetraphenylmethane, that is in the sense of joining suitable building blocks, he used triphenylchloromethane as a raw material this time and allowed it to react with silver (everyone was talking about metal organic reactions at that time). He formulates the desired reaction as follows (Gomberg 1900):

$$2\ (C_6H_5)_3CCl + 2\ Ag = (C_6H_5)_3C{\cdot}C(C_6H_5)_3 + 2\ AgCl$$

The dot in the middle of the planned product molecule symbolizes the single bond between the two carbon atoms of the central ethane fragment. As was mentioned, the attempt did not succeed. Over 80 years later the following passage appeared in a standard book of reaction mechanisms in organic chemistry:

> Hexaphenylethane has not, indeed, ever been prepared, and may well be not capable of existing under normal conditions due to the enormous steric crowding that would be present. (Sykes 1986, p. 301)

The dimer that can *theoretically* be readily constructed from two monomers of the triphenylmethyl radical[19] thus does not exist, although it was described in other-

18 As Gomberg used a quadruply substituted nitrated product for the tetraphenylmethane of 1897 as synthetic (!) proof of its existence, he expected a "hexanitro product" for the "perphenylated" ethane. As we know, this was not the result.

19 The figure illustrates the current interpretation of relations of the products created. The triphenylmethyl radical, which cannot be represented as a substance, reacts with atmospheric oxygen unless isolated.

wise reputable textbooks well into the 1960s (cf. Fieser & Fieser 1968, p. 410). Gomberg quickly realized that he had not created the intended product. On the other hand, he found fascinating results. Whereas the yield was low, the reactivity of the product was surprisingly high. For example, it easily added halogens and could be readily and reversibly peroxidized whereby it was seen to change color. The result of the elementary analysis was particularly surprising in the context of the anticipated course of events. Gomberg found 87.93 % carbon and 6.04 % hydrogen, whereas the hexaphenylethane was expected to be 93.83 % carbon and 6.17 % hydrogen. The product behaved differently in solution, from which it could not be isolated, than in the solid state. Only later did it become apparent that the solid product was triphenylmethyl peroxide (which has the empirical formula $(C_6H_5)_6C_2O_2$ and for which Gomberg specifies the composition as 88.03 % carbon and 5.79 % hydrogen).

Paul Walden dedicates his monograph *Chemie der freien Radikale* to Moses Gomberg, "who nearly 25 years ago discovered the first radical triphenylmethyl and since then worked creatively without interruption on expanding the study of trivalent carbon" (Walden 1924[20]). In this book Walden lists ten signs for confirming the presence of the triphenylmethyl radical, among them, aside from those mentioned above, the formation of addition products with various reagents and the good electrical conductivity in sulfur dioxide solution (Walden 1924, pp. 49 – 52). He lists the following three cases as conceivable conditions for existence:

1. Radicals are *isolatable* chemical individuals with all the characteristics of free and persistent molecules, or on the other hand,
2. radicals are *not isolatable* but occur intermediately as unsaturated fragments of molecules that react with one another to immediately combine into more stable chemical structures as end products, or
3. radicals exist only in our imagination, not as species of substances; we use them as theoretical aids in interpreting chemical reactions.
(Walden 1924, p. 3; original author's emphasis)

20 This work is particularly informative from the perspective of the history of chemistry because it was published precisely in a transitional phase in which, first, experimental detection of methyl and ethyl groups had not yet been developed in Friedrich Paneth's (1929 – 1931) research teams and, second, the "electronic" phase of the description of radicals had not yet gained prominence. Yet near the end of his book, Walden names one of the main proponents of "electronism," Gilbert Lewis, with an unusual statement regarding the Gomberg radical of 1916, to whom he attributes the following formula: $(C_6H_5)_3\ C{\cdot}E_{91}$!

In light of this operational arrangement (which by the way can be analogously applied to all chemical entities[21]) it becomes clear that Gomberg, Walden, and, after significant doubts persisting for two decades, the entire scientific community are inclined to accept version 2 and grant triphenylmethyl the status of an existing (in a limited sense) chemical entity. Yet in any case it is important to realize that both Gomberg and Walden do not understand "radical" in the sense of the modern "electronic" definition mentioned at the beginning of this section but the "trivalence" of carbon (or similar "deficient valences" with other elements). Although Gomberg advocates his thesis of the formation of a dimer unchallenged as late as 1927–1928 (see the following reaction in the original notation), he explicitly refers to "trivalent" carbon (Gomberg 1928, p. 163):

$$(R_3C\text{-}R_3C) = R_3C + CR_3$$

This interpretation, unconfirmed at the time and now shown to be false, is surprisingly still found in modern outlines.[22] In his conclusion, Gomberg is downright exuberant:

> Hitherto, the definitely recognized units in chemistry have been: atoms, molecules, ions, and electrons. And now, in addition to these four, chemistry, it seems, has to take into account a fifth entity – free radicals. (Gomberg 1928, p. 164)

This lofty thesis from a meritorious veteran of their guild certainly struck a sympathetic chord with the audience of the Chandler lecture at Columbia University as it still does with the inclined readership today. This statement by a synthesist is very informative from the perspective of the history and philosophy of science as well. First, it shows that even this chemist, who for years has been engaged in preparative laboratory work more or less exclusively, now swears an oath to atomism. While this may not surprise some readers, I may remind them that his entire argumentation for the presence of triphenylmethyl was deduced from the reactivity of its reaction products, not from measurements in atomic physics. Second, it is interesting that Gomberg takes note of the electrons, which have non-substance character (and thus from a practical perspective are theoretical entities) but makes no mention of their significance for chemistry, something that has long been discussed (cf. the odd molecules in Lewis 1923). Third, in connection

21 The system Walden describes resembles the "triadic theory of science" proposed by Rom Harré in which three areas of decreasing observability or, respectively, increasing theoretical abstraction are identified for scientific entities (Harré 1986, especially pp. 70–78).
22 Cf. Perkins (2000).

with the previous point it is questionable whether the thesis of the fifth entity "radical" as a basic unit of chemistry is really tenable.

Gomberg's achievement primarily consists in overcoming the prevailing late 19th-century dogma of the ability of carbon atoms to form four bonds and only secondarily in initiating modern chemistry with radicals. Here I would like to note that the "first free radical" resulted from the pursuit of a *synthetic* research program. The product originally envisioned could not be detected, that is to say the "theoretical design of its substance identity and its possibility of realization in the sense of a guide for synthesis" (Schummer 1994, p. 48) failed. Nonetheless, it is expedient to follow Schummer's suggestion and also, or *particularly,* for cases like the present one to speak of an *experimentalist realism.*[23] This is a form of entity realism, for the existence of triphenylmethyl is demonstrated by what may be termed circumstantial evidence. All of these indications are based in qualitative and analytic intention on chemical reactions. The Berthelot principle is thus in effect even in cases in which there is no tangible substance product but where chemical entities are characterized by means of preparative methods, that is by classic "wet chemistry."

2.4 Conclusions

The aspiration to know and control the substance environment inevitably leads the representatives of the science of substances to the manufacture, preparation, and purification of specimens of substances, which are then compared and placed in a systematic order. Marcellin Berthelot was the one who best expressed this fact: Chemistry creates its own epistemic objects, in particular by building up (synthetically), but also by breaking down (analytically), in any case by changing. The conventional philosophy (theory) of the experiment has until now limited itself primarily to the analytic confirmation of hypotheses even regarding the chemical sciences. It follows the description that Berthelot also provided for the other experimental sciences: "What they have come upon by thinking, they confirm by the experiment." (Berthelot 1877, p. 294) This entirely overlooks a fundamental characteristic trait of chemistry, namely its manufacturing character. Yet similar thinking also prevails in broad areas of the modern philosophy of chemistry in which a theory-centric discussion has revolved around what chemistry is not. Berthelot makes it clear that the creative force of the science of substance surmounts those of the other sciences in one pivotal aspect,

23 Related realism conceptions are found in Bensaude-Vincent & Simon (2008) and Chang (2012).

> because it penetrates deeper into the essence of the natural bodies and advances to the elements of these. It not only creates phenomena, it is also able to bring forth a number of artificial bodies which are similar to the natural ones and have all of their characteristics. (Berthelot 1877, p. 294)

This creative force was molded into its specific form in the combinatorial variety of the organic chemistry of the 19th century. Yet the metachemical *Berthelot principle* is also effective in cases in which no tangible substance product is producible. It is the basis for a non-naive experimentalist realism in chemistry. The question remains as to why chemists strive to manufacture new substances, often without a specific goal. This question cannot be conclusively answered by the Berthelot principle alone. Berthelot himself indicates that higher epistemological goals may play a role here when in the same passage he adds, referring to the aforementioned synthetic products:

> These artificially exposed bodies are the realized images of the abstract laws of which chemistry strives to acquire knowledge.

Chemistry is customarily denied such epistemological and ideological ideals, especially by those who misunderstand it as a subdivision of a mathematical-physical description of the world.[24]

It is manufacturing success that chemical activities give priority to, not the power of persuasion of the subsequent theoretical permeation. For this reason I have purposely chosen examples in which the principle *avant les lettres*, which is ubiquitous in chemistry, is particularly evident. Naturally, this by no means signifies that theoretical efforts in chemistry are hopeless, rather it only puts them in their proper place.

---

24 One should not exclude creative and playful motivations in this question. Johan Huizinga offers an interesting approach, admittedly without respect to chemical activities, in his book about the origin of culture in play, for example in the chapter "Spiel und Wissen" (Huizinga 1956, pp. 105–117). A statement by Gomberg at the end of his initial publication on triphenylmethyl demonstrates that competition plays a role in the preparative search for new entities: "This work will be continued and I wish to reserve the field for myself." (Gomberg 1900, p. 771)

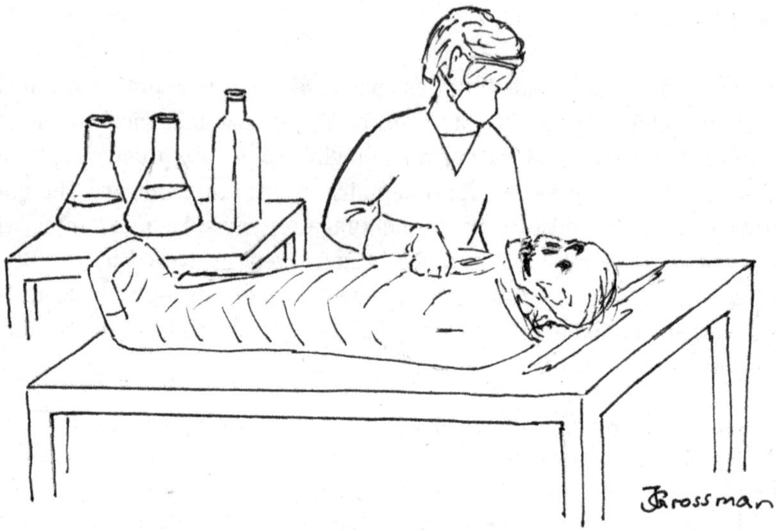

Fig. 4: Pure substances and even elements are preparations – like mummies or skeletons.

3 Chemical Empiriocriticism – František Wald

Many experts in today's natural sciences, especially those in chemistry, cannot imagine that in their disciplines they could make do without atomism, that is without the belief in and discussion of essential submicroscopic particles of matter (or substances). In their (scientific) worldview, the world simply *consists of* atoms and all explanations are reduced to them. Yet chemistry without atoms exists if one only takes the substance concept seriously, examines actual chemical practice without bias, and is willing to exclude what has become highly developed talk of non-observables.[1] The Bohemian chemist František Wald (1861–1930) posed the question of whether it would be theoretically possible to do without the atom model and if so then to what extent. He was among the atom skeptics of the late 19th century, which also included more famous scientists like Marcellin Berthelot, Ernst Mach, Henri Le Chatelier, and Wilhelm Ostwald. Wald assumes an exceptional position in this group because he actually undertook to develop stoichiometry (the theory of the numerical composition of substances) without hypothetical assumptions of atoms. As we will see, in so doing he radicalized the Berthelot principle in a certain sense.

3.1 The Unloved Atomic Hypothesis

In a brief pamphlet entitled *Ein österreichischer J.B. Richter* Wilhelm Ostwald in 1907 writes as follows:

> One generation of chemists after the other was brought up in the belief that the atomic hypothesis answered all questions that could be asked of science with respect to the fundamental stoichiometric law, and a full century had to pass before the thread could be taken up where death had torn it from Jeremias Benjamin Richter's hand. (Ostwald 1910, p. 319)

The so-called "Austrian Richter" Ostwald refers to is none other than the protagonist of this chapter. Ostwald describes the essential result of Wald's studies (as of 1907) as follows:

> In arduous intellectual toil, the grueling nature of which only that person knows who himself has attempted something similar even if not having achieved it, he [Wald] pulled himself upward from one step to the next until he reached complete clarity and realized that the re-

1 As to my previous attempts on Wald confer Ruthenberg (2007a, 2007b, 2008b, 2011, 2012a). I refer to those objects that are not perceivable with the unaided senses as "non-observables." The concept of the "theoretical entity" is closely related.

https://doi.org/10.1515/9783111401041-006

> quired elements from which the law of conservation of mass necessarily results are already contained in the concept of the *pure substance* and the facts that have led to the formation of this concept. (Ostwald 1910, p. 322; original author's emphasis)

Here Ostwald primarily refers (implicitly) to Wald (1895).[2] The proof that the law of definite proportions can be regarded as a metachemical norm is for Ostwald the main result of the "arduous intellectual toil." Ostwald makes no further reference to Wald's writings, for example to those of the third period of his work.

Skepticism about the atom is a conspicuous point of agreement among several pioneers of mathematical chemistry. Aside from Jeremias Benjamin Richter and František Wald the group also includes Oxford chemist Benjamin Collins Brodie (1817–1880) with his *Chemical Calculus*.

3.2 Non-Atomism from Metaphysical Economy

Ernst Mach (1838–1916), an Austro-Hungarian (Bohemian) physicist and philosopher, can be regarded as the main proponent of positivism in the German-speaking countries. Positivists generally regard experience as crucial to the emergence of knowledge and are thus *empiricists*. They play down the role of cause and therefore also that of explanation in the sciences and avoid referring to theoretical entities (*non-observables*) as real objects or phenomena because they are against (supposedly) superfluous metaphysical assumptions.[3] Mach himself prefers to refer to his position as *empiriocriticism.*[4] Yet one cannot reduce him to anti-metaphysics. He spoke of scientific development, that is of *theory dynamics*, and assumed a principle of evolution for it.[5] He saw a further fundamental characteristic of the sciences in the principle of economy of thought, which he described repeatedly, for example in a lecture in Vienna on 25 May 1882. The actual beginnings of science, he states,

---

2 In his *Prinzipien* (cf. Chapter 5), written at about the same time, Ostwald formulates the deduction as follows: *"If two pure substances A and B form a compound pure substance AB, then the proportion of the weights of A and B required to form AB is unchanged within the common range of stability of the three substances."* (Ostwald 1907, p. 350)

3 Cf. Ladyman (2002, p. 148). Other modern introductions to the philosophy of science provide similar characterizations, cf. Hacking (1983, pp. 41–57) and Losee (2001, pp. 143–157).

4 With the polemics directed more toward his domestic political opponent Bogdanov in his widely read primary philosophical work *Materialism and Empiriocriticism*, Lenin (1985) greatly increased awareness of Mach's position yet also greatly impeded critical examination of it, cf. Wolters (1991, pp. IX–XXII).

5 Cf. Haller & Stadler (1988) and Bradley (1971) regarding Mach's philosophy of science.

> only occur in society, and especially in the trades with the necessity of communicating experience. Only there, as many an author has felt, does the compulsion develop to become clearly aware of the important and essential characteristics of an experience for the purpose of description and transfer. What we call instruction merely serves to economize one person's experience by [utilizing] that of another. (Mach 1897, p. 213)

Mach took this notion of economizing or simplifying with respect to study and applied it analogously to the sciences, referring to it as *economy of thought.* He understood this phenomenon, initially observed empirically, as the principle – or norm – of a correct scientific methodology. The principle was then interpreted from the philosophical perspective as *metaphysical economy.* The latter means that in the description and interpretation of the empirically observed situations the "proliferation of the realm of imagination" (the title of a chapter in his book *Erkenntnis und Irrtum*) is to be restricted as much as possible. A passage from this chapter demonstrates this:

> Think of Newton's light particles, Democritus' and Dalton's atoms, the theories of modern chemists, the cage molecules and gyrostatic systems, and finally the modern ions and electrons. The various physical substance hypotheses, the vortices of Descartes and Euler, which are resurrected in the new electromagnetic flow and vortex theories, the sinkholes and springs that lead into the fourth dimension of space, the ultramundane particles that produce gravitation, etc., etc. also warrant mention. I think that is an alarming coven of adventurous modern notions [...] Countless of these flights of fancy and offshoots have to be destroyed by merciless criticism in the face of the facts before one of them can further develop and persists for a while. To do justice to this process, consider that the goal is to reduce natural processes to the simplest conceptual elements. Yet *comprehension* of nature must be preceded by *acquisition* by the imagination in order to create lively illustrative content for the concepts. (Mach 1906, p. 106)

Both the creativity of the researchers and the historical contingency of the respective hypothetical and theoretical statements are thus anything but foreign bodies in Mach's philosophy of science. His own methodological preference was not only the struggle against hypothetical fantasies but above all conceptual economy.

His slightly younger contemporary, fellow physicist, and, initially, atom skeptic Max Planck (1858–1947) also critically examined positivism, albeit significantly later. There are passages in Planck's philosophical writings that indicate that he had by no means rejected the positivist standpoint completely. Upon closer examination, his "realistic" standpoint is found to be reconcilable with Mach's in important aspects. I quote here one such passage to illustrate the disputes about positivism and to further clarify Mach's standpoint. The quotation comes from the lecture *Positivism and the Real Outside World* held in Berlin on 12 November 1930:

> Although the foundation that positivism gives physics is soundly underpinned, it is too narrow; it must be expanded by an addition whose significance consists therein that the science where possible is freed from the contingencies that the reference to single human individuals has brought into it. (Planck 1979, p. 234)

Planck thus expressly concludes that access to scientific (physical) knowledge can only be through sensory experiences. This is the basic premise of positivism. In the second part he demands something exceedingly similar to Mach's position, namely an objectification of scientific statements, which according to Mach should be achieved via metaphysical economy. After the sentence quoted, Planck states that this "objectification" occurs "not by means of formal logic, rather [with the] step into the metaphysical dictated by levelheaded reason" and then postulates – similarly to Kant – a dichotomous worldview: a world of experiences on the one hand and a world that stands behind it. Like many other natural scientists of his time, Planck apparently sees the world behind the experiences in the Kantian tradition (the *noumenon*) as the actual world. Whereas for Mach metaphysics in scientific work represents a (possibly) necessary evil, for Planck it is the constituting and driving force of successful physics. This is the essential difference in the positions of these two physicists regarding the philosophy of science.

Mach's *History and Root of the Principle of the Conservation of Energy* was published in German in 1872. In it he provides an early sketch of his anti-materialist ideas. Within the scope of his criticism of the mechanistic theory of heat he also makes observations about chemical topics:

> To us natural scientists the concept of "soul" is occasionally offensive and we smile at it. Yet substance is an abstraction of exactly the same sort, as good and as bad as the first one. We know as much about the soul as we know about substance. (Mach 1872, p. 25)

In the same passage Mach describes the *dispositional* behavior of chemical substances, without using the term, as follows: When oxygen and hydrogen explode, "then the oxygen and hydrogen phenomena disappear and water phenomena occur instead." Mach regards the assertion that water consists of oxygen and hydrogen as a linguistic stop-gap solution. Comparing the heat substance and substance concepts, he states:

> Heat is a substance just as much as oxygen is one, and it is not one as oxygen is not one. Substance is a possible phenomenon, a fitting word for a hiatus in thinking. (Mach 1872, p. 25)

Describing the term "substance" as a "fitting word for a hiatus in thinking" is a brilliant piece of positivist polemics. As we will see below, Wald, too, was capable of such expressive language. In any case Mach discusses in the same passage the peculiar aspect of the latent chemical elements, that is the phenomenon that in certain operations previously undetectable basic substances (re)appear; think of the electrolysis of water with the release of the two elements oxygen and hydrogen. Thus we may credit Mach with recognizing and clearly formulating this fundamental epistemological problem, one which to this day is still present in the discussion of philosophy of chemistry.

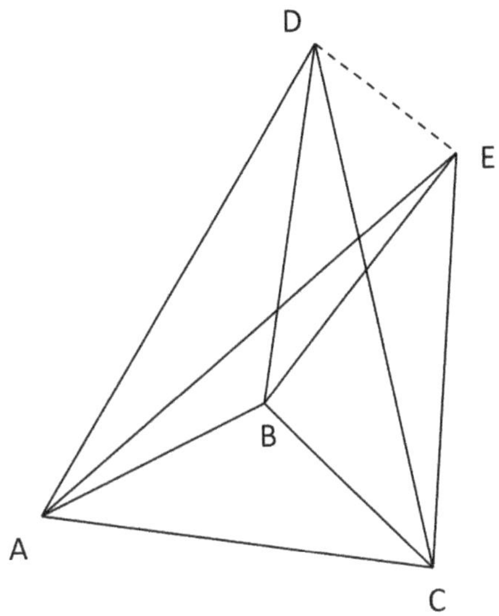

Fig. 5: A virtual 5-atom molecule according to Mach (1872, p. 29).

Yet what was František Wald referring to in the *Annalen der Naturphilosophie* with his discussion of multidimensional geometry in connection with chemical theories? Taking an entirely sensualist stance, Mach states that it is an unnecessary limitation for the scientific process "to conceive of that which is merely thought as spatial, i. e. with the relations of the visible and tangible," and he compares these forms of sensory perception with the acoustic form, which if one were to think of a certain tone pitch would be just such a limitation. Here one could presume multidimensionality in the sense of a "multisensual" examination. Yet that is not the sort of multidimensionality that Wald and Mach had in mind. Rather the Bohemian chemist was thinking of a hypothetical five-atom molecule. With this idea (Figure 5), Mach wants to show how limitation to three-dimensional space can lead to limitation of the possibilities for describing chemical phenomena. If one begins with the tetrahedron ABCD and adds atom E – here Mach actually uses the words "atom" and "molecule"–with defined distances to each A, B, and C, then the distance DE is also determined. It follows that DE cannot be independently changed. This has the following consequence:

> There may well be serious difficulties in imagining several 5-atom isomeric molecules that differ only in the relationship of D and E. This difficulty disappears in our example when we imagine the 5-atom molecule in 4-dimensional space. (Mach 1872, p. 29)

Mach sees this as an example of the inability of simple formula contrivances to adequately represent the full range of conceivable relationships between elements. Here I will refrain from critically examining the applicability of the chosen example at length; for example the basic premise chosen here, that all atoms in a molecule directly interact with one another, was already questionable in the chemistry of that time. Here it is sufficient to have identified the site that Wald was specifically referring to when he spoke of multidimensional geometry and from which he, as we will see below, took his methodological inspiration. Wald repeatedly emphasized that his own methodology followed "Mach's process."[6]

3.3 Chemical Positivism in Prague

In this section I will undertake the attempt to critically appraise František Wald's writings in the *Annalen der Naturphilosophie*, i. e., of writings from the middle and late phases of his work in the theory of chemistry.[7] I will examine those essays and passages of his contributions to the *Annalen* from 1902 to 1909 that I feel are crucial to illustrating this approach, one unique in the history of science. Of the authors contributing to the *Annalen der Naturphilosophie*, František Wald, with a total of nine essays in the years 1902–1909, published more than double the num-

6 Cf. Wald (2004, p. 223).

7 In the first phase, which extended from his initial publishing activity in 1881 to the early 1890s, Wald examined basic issues of thermodynamics and, among other things, described entropy as devalued energy (Wald 1889). In the late middle phase in which the articles in the *Annalen* were written, he proceeded from Gibbs' phase theory and attempted to reconstruct chemical stoichiometry in phenomenalistic terms. He noticed that the premise of preexisting substance components as assumed in Gibbs' phase rule ($P + F = B + 2$, where P is the number of phases present, F the number of degrees of freedom, and B the number of components) represented a fundamental inconsistency of his approach. This led him in a third period of work to attempt to move away from the phase rule and, additionally, from the naively realistic assumption of pure substances (elements, compounds). Now he wanted to find a mathematical geometrical path to reconstruction to represent this and other substances while avoiding a circular theoretical argument. However, he retained the concept of "phase" and used it as a generic concept under which other concepts such as "substance" were subsumed. He gave his main work, which due to the chaos of war was published in 1918 only many years after he had begun writing it and represented the culmination of his theoretical contributions, the title *Chemie fasí* [Chemistry of the Phases]. Wald's third period of work began while he was writing the essays for the *Annalen* and ended with his death.

ber of articles than did his chemist colleague (and co-editor of the journal) Wilhelm Ostwald. And yet as evidenced by a letter to Ostwald of 28 June 1901, he initially reacted very reservedly, almost negatively, to Ostwald's suggestion to publish his articles in the *Annalen* in the future:

> Your messages about the new journal of natural philosophy delighted me […] However, I still have a certain reservation; the epistemological part of my work is not its actual goal; I have only done these studies because I had to clear the way for myself. (Pinkava 1987, p. 105)

Wald thus regards himself as more of a theorist of chemistry. Ostwald had asked the Bohemian two days previously to publish his article not in *Zeitschrift für Physikalische Chemie*, but in his new *Annalen*. Thus he unintentionally contributed to (further) isolating Wald and his theory in the scientific community of chemists. For with very few exceptions hardly anybody in the field of scientific chemistry gave serious consideration to Wald's approach. The response to Wald's articles, hardly exuberant even before 1902, dried up nearly entirely afterward.[8] Not only did Wald and Mach both live in Bohemia and share the more or less coincidental fact of being among the first very authors of the *Annalen der Naturphilosophie.*[9] The two of them had known each other for over ten years when the first volume was published, initially via correspondence[10] and later personally as well. Mach was a professor of physics at the University of Prague from 1867 to 1895; Wald studied chemistry at the Technical University of Prague until 1891 and later worked as the director of the chemical laboratory at the steelworks in Kladno, hardly 30 km west of Prague.[11] Wald could easily have attended lectures presented by the well-known physicist and philosopher (there was no genuine student-teacher relationship). In any case the two natural scientists were in close contact. By his own admission, Wald came into closer contact with Mach's thinking, particularly the application of multidimensional geometry to chemistry, when he studied Mach's *History and Root of the Principle of the Conservation of Energy:*

8 A notable exception is Richard Ehrenfeld's (1877–1909) introduction to the history of chemistry. He states: "The proposed change in the understanding of the chemical individual, which simultaneously signifies a wholehearted emancipation from the previous atomist views, gives rise to a number of new questions […]." Ehrenfeld (1906, pp. 271–287)

9 Machs article *Ähnlichkeit und Analogie als Leitmotiv der Forschung* after the editorial by Ostwald is the first published article (reprinted in Mach 1906). Wald's abridged version of *Kritische Studie über die wichtigsten chemischen Grundbegriffe* (Wald 1902a) is the second.

10 Cf. Thiele (1973) and Pinkava (1987).

11 For biographical and bibliographical details cf. Druce (1944), Pinkava (1987), Ruthenberg & Psarros (1994), and Ruthenberg (2007a).

> My attention was drawn to multidimensional geometry by a comment from Prof. *Mach*, who had recommended over thirty years ago that chemists study it; I read the comment around the year 1890 in an article about the root of the principle of the conservation of energy and since then have adhered to it as well as I was able. (Wald 1909, p. 220)

Wald's first book *Die Energie und ihre Entwertung* was published in 1889. In it he undertakes a phenomenalist interpretation of the law of entropy on the basis of the following assessment:

> One has taken great pains to deduce the law of entropy from molecular motions, and that this undertaking has succeeded is regarded as particularly valid proof of its correctness. In my opinion, the roots of this law lie far deeper, and successfully having brought the molecular hypothesis in line with the law of entropy is fortunate for the hypothesis but not for the law of entropy. (Wald 1889, p. 104)

The statement made by this second sentence has a history of reception. Wald's assessment and Ernst Mach's aforementioned study were favorably received. Mach quoted him approvingly in his "Principien der Wärmelehre" with the explanation that there could be no increase in entropy in a "purely mechanical system of absolutely elastic atoms," and therefore the molecular approach by scientists such as Ludwig Boltzmann (and later Max Planck) to "explain" the second law of thermodynamics represents an artificial correlation. The response of those researchers who like Boltzmann did not want to abandon the atomist-molecularist realm of argumentation (and maybe even worldview) was naturally less friendly also because in their view the description of distributions of particles in the form of "microstates" was not understood as metaphysical speculation. Wald had thus applied Mach's methodology or at least something very similar to it even before reading *History and Root of the Principle of the Conservation of Energy*. Although in this first thermodynamic period of his work, he utilizes the vocabulary of particle-centric and compositionist chemistry and refers to *preexisting* elements (he later calls this view *synthetic*),[12] this period can be regarded as the formation phase of his positivist views. In his main work *Chemie fasí* he describes the transition to these views in one passage as follows:

> Around 1893 I resolved to review the whole of previous chemical thinking in order to rid it of hypothetical notions and to emphasize therein, according to the process of *Mach*, only *chemical experiences in phenomenological representation*. (Wald 1918, p. 223)

12 His Czech-language lecture on dissociation theory contains little that might indicate an anti-atomist or non-atomist. Cf. the English translation of this lecture in Ruthenberg (2009, pp. 13–23).

Upon closer examination of the text sources we can proceed from the assumption that even before his close acquaintance with Mach and his works František Wald had assumed what we would call in modern terminology an *anti-realistic* attitude. This independent intellectual attitude became even more apparent later with respect to Wilhelm Ostwald. Here it will be the Bohemian's influence on the great Leipzig researcher that stimulates the latter to work in a related sector and not the other way around.[13]

Wald soon recognized that his earlier attempts to deduce the stoichiometric relationships in compounds with algebraic equation systems was a deficient representation of existing knowledge whose content did not go beyond contemporary atomist-compositionist interpretations.[14] After a few years he changed to a principle from which he removed the phase rule with its pure constituents and used only the phase concept. This pivot took place in the phase of his articles for the *Annalen der Naturphilosophie.*

3.4 From Skeletons and Mummies to Multidimensional Stoichiometry

Wald published three essays in the first volume of the *Annalen*, the first of which was a concise overview of the second (both were titled *Kritische Studie über die wichtigsten chemischen Grundbegriffe*) and the third of which was a supplement to the first two titled *Über einen alten Denkfehler in der Chemie.*[15] In the concise overview of his "critical study" he maintains that the emergence of certain lines of inquiry in mixed-phase thermodynamics in physical chemistry had led to a situation where the uncritical acceptance of the atomic hypothesis and the "structural doctrines" had to be reexamined.[16] He goes on to state:

> Whereas chemical elements and compounds are customarily regarded as finished building blocks of natural substances, as products of an act of creation, unbiased observation of the process of their development reveals them to be *artificial products*, the result of a series of

---

13 Cf. Ruthenberg (2008).

14 Cf. Pinkava (1987, p. 18). Also the basic statement of the law of mass action, that no chemical reaction proceeds completely, is not representable with this approach.

15 Cf. Wald (1902 a–c).

16 It is useful to bear in mind that until elementary particle physics and quantum mechanics became fully developed in the second quarter of the 20th century, the "atomic hypothesis" was usually understood to mean the notion of submicroscopic inert material particles behaving in a purely mechanical manner. Current notions of "atom" have become far removed from this mechanical image and are significantly more complex.

> painstaking chemical operations, that is to say *specimens*, much like a skeleton or a mummy. (Wald 1902a, p. 16; original author's emphasis)

This representation, as extreme as it is essential for Wald, was at the time and still remains a massive provocation for the chemical community. The chemists of the late 19th century had already become accustomed to accepting the standpoint Wald depicts as naively realistic, that is to believe in a material substrate, and some had wildly speculated about the characteristics of the building blocks of this substrate. Many specialists in the field harbor this belief to this day – together with a good number of contemporary philosophers of chemistry. How it might be possible to penetrate down to the elements is a question that hardly anybody seems to regard as epistemologically relevant. The chemical elements as preparations (Figure 4): Nobody has ever attacked the basic tenets of the chemical sciences so radically, and nobody has ever taken the positivist-operationalist methodology so seriously. Based on the fundamental empiricist notion of preparation, Wald views all homogeneous substances initially "as completely equivalent entities." He initially places substances of defined composition such as compounds and elements and substances of variable composition on the same hierarchic and systematic level, as the empirical proof of the existence of an element is the *result* of the operational interaction with the substance environment and not its point of departure. He makes another outrageous statement in the same passage in which he describes his emerging doubts about "whether when observing every substance we are really compelled to ask about its chemical *composition*" (Wald 1902a, p. 17).

If one regards chemistry in Wald's sense as the scientific examination of the phases and their transformations, then compositionist examinations are only part of its methodological toolbox.[17] Yet if chemistry is understood primarily or exclusively as the science of the *compositions* of substances, that is as a breaking-down-and-analyzing undertaking in *Lavoisier's sense*, then Wald's concept must be understood as an attack on the very foundation of the scientific undertaking so characterized. A more phenomenalist definition (as the study of substances and their transformations) leaves open *how* substances are characterized. From this standpoint it would conceivable that the reactivity of a substance of interest could be detected by its chemical behavior, that is by examining a series of its reactions. This does not necessarily require the operations of chemical analysis in the sense of "breaking down." Consequently, the skeptical question quoted will only disturb those who unquestioningly view *analysis* (breaking down) as a necessary

17 Cf. Ostwald's (1894, p. V) eloquent general characterization of analytical chemistry as the "art of detecting substances and their components."

precondition for chemical activities. Interestingly, the author of the "critical study" at the time of its publication was himself employed as a chemical analyst in the steel industry and therefore performed professional compositionist analytics himself.

However in the short version of the "critical study," František Wald goes even further in his demands for a new theory of chemistry. He criticizes the abstracting representation of the formation of reaction products and calls it "superficial" because there is no mention of whether mixtures are created or which separation operations are required. The author continues:

> This *idealization* of reality is useful as long as the intention is merely preliminary orientation. Yet it becomes an impediment to any more precise research because right from the start one declares a wealth of detailed phenomena to be nonexistent, making it subsequently impossible to grasp their existence. (Wald 1902a, p. 18; original author's emphasis)

In contrast to this, Wald proposes a strictly operational description of chemical processes. This view, which he finds preferable to the existing theory, he explains in greater detail in his *Kritische Studie* (Wald 1902b). Here one finds a more detailed characterization of the substance concept, which he derives from the concept of the chemical body. Accordingly, in the case of chemically interesting phenomena, chemical bodies are in direct contact, whereby their size and shape are unimportant. Wald explicitly describes "substance" as an abstraction of "body." In the same context he repeats his theses from the short version, especially the dispensability of the element concept in chemistry and the necessity of physical contact in the case of chemical processes. Here he mentions catalysis and describes the previous image of substance transformations as incomplete because in catalytic effects the substances are in contact but the catalyst emerges from the process unchanged. Its effect is thus not adequately represented in a chemical equation because it appears in this equation as a secondary parameter at best, not as a participant in the reaction. This leads to the contradiction that the presence of the catalyst, which in some cases is actually indispensable, does seem to be necessary according to this representation of the course of the reaction.[18] Wald comments on this:

18 In chemical thermodynamics the reaction energy as a so-called state value in equilibrium is actually independent of the course and the speed of the observed reaction, whereas the catalyst in independent observations for, example via the Arrhenius equation, is observed quantitatively in a certain sense (cf. the discussions of equilibrium and catalysis further below). Wald takes a more general standpoint here and focuses on the overall course of the reaction.

> However, over the course of time this abstraction, too, has proven to be not consistently applicable because it became apparent that the nature of the solvent (even when its quantity does not incur any change by the reaction) indeed often even traces of certain bodies have an extraordinary influence on the course of the phenomenon. (Wald 1902b, p. 198)

Wald correctly notes that the environment of a chemical reaction has a significant effect on its course and describes how contemporary chemical theorists treat new experiences such as catalysis as a "grafting" of additional assumptions by "abstraction" and "schematizing" and thus as a deviation from the "way to any natural observation." He even goes as far as to publish the following contention:

> It does not seem to me at all audacious to maintain that the atomic hypothesis in chemistry currently serves only to correct the superficial observations with which every chemistry class and every chemistry textbook is introduced. (Wald 1902b, p. 212)

These "superficial observations" may still be found in the chemical sciences. Textbooks often present a naively realistic image of the theoretical entities of the natural sciences, whereby closer inspection reveals that not only has the (conceptual) focus of research and study shifted into the submicroscopic realm but the implicitly conveyed ontology as well. Efflorescent definitions of chemistry as the *study of the behavior of the outer electrons* are hardly rare in these texts.

Wald's next major article in the *Annalen* is *Über die Mannigfaltigkeit chemischer Erscheinungen* (Wald 1903). In this reflection on the dimensionality of the geometric representation of interacting state values (pressure, temperature, substance quantity, volume) he explicitly states what will define the methodological approach in the late phase of his work:

> The relation of *one* of these dependent characteristics to *one* of the independent ones is always representable in one plane; if one wanted to represent two characteristics in their dependence on temperature, then an image space of three dimensions would be required; if one intended to depict, say, seven dependent characteristics simultaneously, then one would have to have a space with eight dimensions or – because there is no geometrical space of this sort – break the representation down into a series of planes or (three-dimensional) spatial images. (Wald 1903, p. 109; original author's emphasis)

Wald describes the number of effective independent variables in a process under observation as a "manifold." He laments the absence of this concept in chemistry because with the variety of impressions and experiences that it offers, it seems to him to be nothing less than predestined for his applications. If one disregards the number of degrees of freedom in Gibbs' phase rule, this concept has not gained acceptance in the chemical sciences.[19] Wald's essay reveals "his" philosophy of science. It is a plea for phenomenalism and for metaphysical economy even when in chemistry it is often a matter of variety. In a certain sense it also reflects a methodological plurality that is typical of chemistry. Another unusual step in Wald's early essays is the assumption of the elements as independent variables. He argues for the subordinate position of the chemical elements (at the time about 70 were known) in a strictly operationalist chemistry, citing among other things the fact that experiments can be performed in a chemical laboratory without real samples of these elements. As he is interested in the mathematical side of this field, he estimates the order of magnitude of the number of necessary non-elementary original substances ("root bodies"; in *Chemie fasí* he will then call them "guide members") for a series of chemical reactions and comes to the conclusion that:

> the number of independent weight relationships among the entirety of chemically representable bodies must be incomparably smaller than the previous theory with all its supplements (atomic hypothesis, valence theory, ion theory, distribution laws, laws of mass action, and the like) has been able to detect. (Wald 1902b, p. 120)

In addition to this hint to an argument for simplifying chemical representations, we also receive a list of the main theoretical points of attack from the standpoint of a phenomenalist theory of chemistry. Wald has only words of irony for isomerism as the different spatial orientation of atoms in molecules of the same composition (the same empirical formula). He feels it is inadmissible metaphysical speculation to speak of the spatial orientation of non-observable submicroscopic entities as was then commonplace in mainstream chemistry. As we now know, however, some of the original speculations, although by no means all of them, have proven to be tenable theses as is the case with isomerism.

Although the field of mathematical chemistry only developed into an independent discipline in the second half of the 20th century, it had its historical precursors as have been described; and in any case we must count Wald among these pioneers at the latest with his *Mathematische Beschreibung chemischer Vorgänge* (Wald 1909). This, his last contribution to the *Annalen* – also his first as professor of the Czech Technical University of Prague and his last in German (disregarding

---

19 The article by chemist Gustav Hüttig (1914) in the *Annalen* indicates the popularity of the interpretation of this concept. Hüttig, despite adhering to what he calls an energetic and anti-realistic standpoint, attempts the feat of explaining the constitution of molecules with a manifold concept interpreted merely as "diversity." He mentions Wald once briefly and rather cryptically. I am unaware of whether Wald took notice of Hüttig's article.

his son's German translation of *Chemie fasí* which was not widely publicized and was published much later)–is announced as a "First Treatise."

No further directly related treatise exists, not even in Czech, and one can only speculate as to the reasons for this. Shortly after he became a professor in Prague, Wald's wife died and he was solely responsible for his five children. When he was named professor he was already very exhausted, and the strain of the new and unfamiliar routine of university teaching did not improve things. Finally, an additional aspect related to his personal relationship with Ostwald should be considered. In a letter of 8 July 1912, three years after the publication of *Mathematische Beschreibung chemischer Vorgänge*, Wald writes to Ostwald:

> [...] since my last publication in the *Annalen für Naturphilosophie*, I have conducted a lengthy and arduous refinement of multidimensional geometry for my purposes and would be happy if I could briefly develop its central idea for you and could also show you what the method is capable of achieving. (Pinkava 1987, p. 138)

The desired exchange with Ostwald never materialized. It seems that Ostwald was satisfied with Wald's (metachemical) result established years previously concerning the derivation of the law of definite proportions independently of experience. The work mentioned in the letter represents the material for *Chemie fasí*, which is *de facto* the "second treatise" although it was not referred to as such. Although the latter would follow at a later date, the monograph-like essay of 1909 spanning 52 printed pages represents in many ways the high point of František Wald's work. First, it stands out from Wald's other texts in its structure and clarity. Second, its thoroughness fulfills nearly all wishes. Third, here for the first time the geometric approaches are presented in examples (unfortunately examples are lacking in *Chemie fasí*). Fourth, although it contains some repetition from earlier writings, one can read this "First Treatise" as an introduction to *Chemie fasí* or, in other words, *Chemie fasí* is hardly comprehensible without the "First Treatise" in the *Annalen.* I will shortly examine here an aspect handled in Section 6 of this last essay in the *Annalen* which serves to illustrate the connection to Mach on the one hand while on the other hand foreshadowing *Chemie fasí.* Wald describes the changes in chemical systems depending on the mass of the components (which he still assumes as known, that is preexisting). He uses the example of the gas mixture oxygen, hydrogen, and water vapor.

The mass proportions of these three substances can be represented in a Cartesian coordinate system as shown in Figure 6. Between points o, w, and h – the points for each of the pure substances – lies a plane that Wald refers to as the *Lavoisier space.* Thus he describes the point range formed by the sum of the mass proportions of the substances in question projected onto the coordinate axes. This

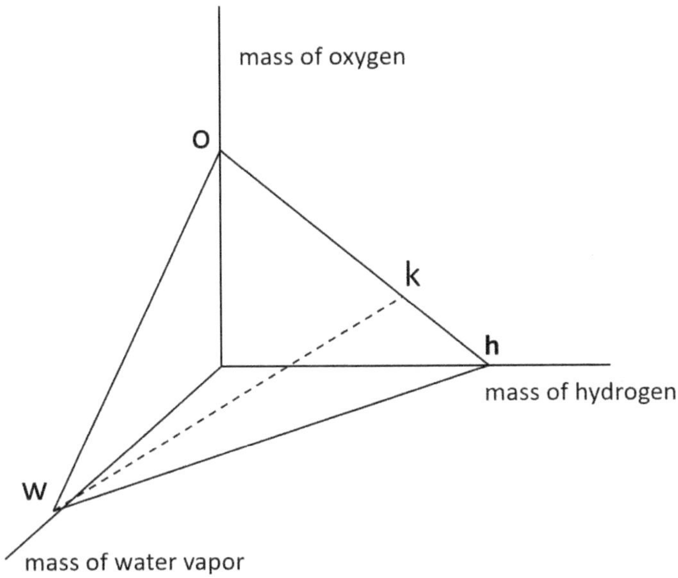

Fig. 6: Cartesian representation of the mass proportions in a gaseous system of three substances oxygen, hydrogen, and water vapor according to Wald (1909, p. 229, Fig. 7). The transposition path k-w describes the transition from "pure" oxyhydrogen gas into "pure" water vapor.

range is two-dimensional for three participating substances, that is a plane. The chemical reactions take place within this plane, in Figure 6 for example along the line k-w: A "stoichiometric" mixture of the original elements leads to pure water vapor.

Such chemical transpositions thus transform existing manifolds. By adding or subtracting certain quantities of the original substances one obtains parallel shifts in the "transposition paths." With this and the geometric methods developed from it, Wald attempts to eliminate the flaw in the existing theory of preexisting elements, which for him is that, with respect to the number of dimensions, it stops at the manifold of the elements assumed in advance, from which the following conclusion can be drawn:

> It [the prevailing theory] then cannot admit any formation and any disappearance of the elements and must replace the missing dimensions in another way; for example it assumes a "rearrangement" of the atoms in space. (Wald 1909, p. 232)

At this point the difficulties of making the concept of substance compatible with a submicroscopic mechanical description of the world again become apparent. To

Wald's mind, the two conceptual realms were even incompatible. In place of a plane as a Lavoisier space as in the previous example, the quadripartite reaction A + B → C + D results in a three-dimensional structure (Figure 7). The transposition

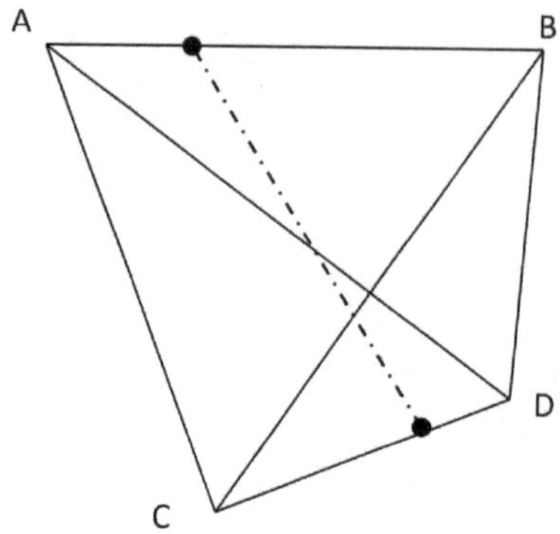

Fig. 7: Geometric representation of the three-dimensional Lavoisier space for reactions between four components according to Wald (1909, p. 233, Fig. 12). Similarly in Wald (1918, p. 164, Fig. 15).

path for this reaction passes through the tetrahedron from the edge AB to the edge CD.

The use of multidimensional geometry as in these chemical examples has no specific historical precursor, but in conceptual terms it can undoubtedly be traced back to Mach. Wald arrives at quasi multidimensional representations in the later period of his work. In so doing his focus is on the systematic elucidation of the operational path of chemistry to the state of knowledge of that time. His choice of words is unusual. For example, he describes the contemporary atomist-compositionist theory of chemistry as *synthetic:*

> We can briefly describe the prevailing chemical theory as *"synthetic."* It teaches us the *composition* of some substance from elements assumed as given. It leaves the structure of the other phases to chemical practice and physical chemistry and regards this task as fruitless for pure chemistry. (Wald 1918, p. 130; original author's emphasis)

I find the description "synthetic" for the compositional concept of chemistry a poor choice of words as precisely the part of chemistry that Wald wants to emphasize and which he in contrast terms "analytical" is in fact far closer to the manufacturing and preparative character of chemistry that he himself emphasizes. He also calls his approach *a priori* (Wald 1918, p. 199) in the sense of "before experience":

> I want to follow the train of thought of a chemist who neither knows anything of elements nor of compounds into which he can later separate his working material. His raw materials and his reaction products are phases of unknown random composition of unknown elements and unknown compounds. (Wald 1918, p. 130)

Wald projects a powerful metaphor for the core of his line of inquiry in the science fiction novel *From the Earth to the Moon* by Jules Verne.[20] Assuming that one would find oneself in an environment of completely unknown substances, it would be interesting to know how to proceed in terms of method to determine which materials are present and whether or not the substances found are similar to those on Earth. At the time the novel was written and even at the end of the 19th century it was not at all clear whether the same stock of substances would be found on the Moon as on Earth. The fundamental methodological question for the historical reconstruction here on Earth is similar to that which travelers to a distant heavenly body would pose:

> *How must the phases in general behave if in their application the doctrine of elementary substances and their compounds could arise which no longer recalls the phases?* (Wald 1918, p. 131; original author's emphasis)

Wald rightly criticizes the attitude of his contemporaries from chemistry who were inclined to treat chemical matters without hesitation from the standpoint of the prevailing theoretical interpretation and in so doing unquestioningly subordinated the (historical) "how?" to the (current) "what?"

3.5 Conclusions

František Wald regards the phenomenologically easily accessible phase concept as the starting point of all chemical endeavors and sees the substance concept as an abstraction of that concept. Contrary to the attitude prevailing since the chemical revolution that an element concept is necessary for interpreting chemical phenomena, Wald sees the modern elements, like all other "pure" chemical substances, as the result of preparations. Compositionism, which since Lavoisier at the latest developed into a fundamental dogma of the science of substances, is in Wald's theory of chemistry consequently only one possible form of studying substances.

If one considers his personal background, Wald resembles a "single tree standing outside the forest," a *solitary figure.* He lacks the deeper academic background

20 Cf. Verne (1865). In this early science fiction novel, a team of adventurers is shot toward the Moon in a large projectile by a gigantic cannon. The projectile misses the Moon but remains in its orbit. In the sequel to the novel, the protagonists successfully return to Earth. Jules Verne makes no mention of the situation of the chemical examinations on the surface of the Moon required for Wald's thought experiment. Wald mentions his Moon metaphor in Wald (1895, 1897, 1909b, 1918).

Fig. 8: How to begin chemistry on the moon?

and the collective corrective it provides; despite good results in his studies he remains without a formal degree (which his industrial career did not require) and therefore without any additional live connection to a professorial advisor or working group. His correspondence shows that even early on he endeavored to achieve something special. He is not a classic academic, tends to read little, and is quickly ready to fight for an idea once he accepts it as correct. He also works for many years in the analytical quality control of metal materials in a steel plant. His professional *grounding* has undoubtedly had a decisive influence on his material-oriented theory of chemistry, a field of his own choosing that he pioneered, yet one that remained greatly limited in scope.

In a certain sense Wald radicalized the Berthelot principle. Despite its very plausible metachemical starting point, his multidimensional reconstruction of the characterization of substances is too cumbersome and one-sided. Due to his self-imposed limitation to stoichiometry, it necessarily leads to merely another form of compositionism. Yet his preparation doctrine as the fundamental idea of the philosophy of chemistry nonetheless remains exceptionally valuable. This doctrine is very close to what the French philosopher Gaston Bachelard (1884–

1962) said a few decades later, when he explicitly introduced the concept of metachemistry, and which Bernadette Bensaude-Vincent comments as follows:

> [...] metachemistry takes into account all the technical work necessary for chemical substances to exist: they are real because they have been 'realized,' that is, synthesized or purified. (Bensaude-Vincent 2012, p. 145)

In that perspective, František Wald is a pioneer of metachemistry. His doctrine makes it clear that a chemistry without atoms is not only possible but that it already exists in practice and has existed since humans have actively and systematically interacted with their substance environment. A large part of actual practical chemistry is non-atomic.

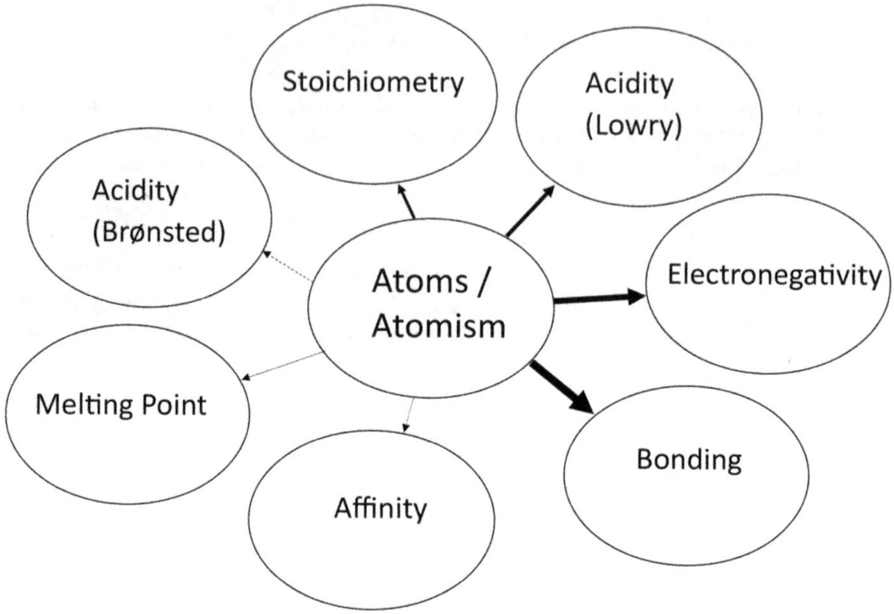

Fig. 9: Selected chemical measurement variables, terms, and concepts in connection with fundamental atomistic ideas. The utility and/or influence of (chemical) atomism increases as we move clockwise from the lower left (affinity).

4 Foundations of Chemistry – Ostwald's Principles

Without a doubt Wilhelm Ostwald (1853–1932) assumes a salient position in both the history and philosophy of chemistry.[1] Many standard relationships in physical chemistry reflect his influence and some even bear his name to this day: Ostwald's law of dilution, the Ostwald process (the catalytic combustion of ammonia to produce nitric acid), chemical thermodynamics and chemical kinetics, and the ideas of catalysis (Nobel Prize 1909) are particularly famous. Ostwald also acted as an intermediary and multiplier in other, more familiar areas of the history of philosophy: He published the first German edition of the *Logical-Philosophical Treatise* by the young Ludwig Wittgenstein in his journal *Annalen der Naturphilosophie* in 1921. In spite of his fame and his far-reaching scientific achievements and influence, philosophers of chemistry have written little about his works.[2] Therefore in this chapter I will undertake an attempt to critically appraise his 1907 work[3] *Prinzipien der Chemie* [Principles of Chemistry], a work that received particularly little attention but is otherwise of great interest to the perspective chosen here. Here I will focus especially on the concept of substance and on reactivity.

4.1 An Introduction to All Chemistry Textbooks

Principles of Chemistry, from whose subtitle I have taken this section heading, is both an extensive (540 pages) and unusual introduction: Because he rejected the atom concept, this book lacks any formulaic representation, meaning that even in the section *Chemical Formulas* (Ostwald 1907, pp. 393–401). Ostwald himself uses neither empirical nor structural formulas. One rather strange exception is a figure that depicts three isomeric variations of a *virtual* compound A_3B_2 to ex-

1 Cf. Simon et al. (1982), Servos (1990), Laidler (1993), Schummer (1996).

2 The few available studies include Mittasch (1951), Bradley (1955), Simon et al. (1982), Niedersen et al. (1992), and Görs et al. (2005). Even the entry about Ostwald in a recent handbook on the philosophy of chemistry refers exclusively to his energetics and makes no reference to his *Principles of Chemistry*; see Deltete (2012).

3 Keith Laidler, in his otherwise quite comprehensive and enlightening book about the history of physical chemistry *The World of Physical Chemistry*, does not even mention *Principles* in the bibliography (Laidler 1993). The same thing occurs in the proceedings of a conference devoted exclusively to Ostwald (Görs et al. 2005).

https://doi.org/10.1515/9783111401041-007

plain the concept of valence (Ostwald 1907, p. 510).[4] Consequently, no reaction equations may be found as the book surprisingly manages to do without almost any examples from practical chemistry.[5] Yet it uses a wealth of one- and two-dimensional phase depictions and conventional phase diagrams. Unfortunately, it also lacks the customary source citations and bibliographic information.

Principles consists of 11 chapters that largely build on each other in a logical sequence: First the concepts of bodies, substances, aggregate states, solutions, and equilibriums are introduced (Chapters I–V); then it continues with elements and compounds (VI), stoichiometry (VII) and colligative characteristics (VIII); then on to kinetics (IX), isomerism (X), and finally to ions (XI). The longest chapters are *V Solutions* and *VI Elements and Compounds* with about 100 and 110 pages respectively. The latter is the most important in terms of the significance of its content in the current context.

4.1.1 The Motivation

Why does Ostwald write this book? At the time of its publication he is no longer a professor at the University of Leipzig, where he had headed the world's first department of physical chemistry since 1887, was eminently successful, and had trained an enormous group of physical chemists, especially from other parts of the world (Servos 1990). Bear in mind that Ostwald, at the latest since assuming the chair of the department at Leipzig, had developed his *energetic* view in which *energy* was to be given the highest priority in the observation and examination of natural phenomena and *matter* was subordinate to it.[6] From this perspective and under the influence of Ernst Mach, Ostwald, after an initial phase as an atomist, soon developed a fundamentally non-atomist attitude. In his atom skepticism and particularly in his phase-oriented operationalism, Ostwald was significantly influenced by Wald, as is made clear by the text of his Faraday lecture of 19 April 1904:

4 However, these do not really make sense without an atomistic interpretation as Bradley (1955, p. 36) shows.

5 Examples of exceptions in the specification of real characteristics or substances are (with page numbers from the *Principles*): coefficient of thermal expansion pp. 48–49, density of water pp. 58–59, vapor pressure of water pp. 103–104, allotropy of mercury iodide pp. 138 and 144, red and white phosphorus pp. 142–143. Here reference is made to the German original. The English edition, translated by Harry W. Morse, is Ostwald (1909).

6 Ostwald describes the birth of energetics in Berlin's Tiergarten district very vividly in *Lebenslinien* (Ostwald 1927, pp. 160–161).

> Up to the present moment, the question of whether it is possible to deduce the stoichiometrical laws without the help of the atomic hypothesis has only been raised by other investigators in order to deny the possibility. As far as I am aware, there exists only one man who has worked on the question with the earnest hope of obtaining an affirmative answer... The man is Franz Wald... To him I owe first the idea that the definition of substances and elements is in a certain sense arbitrary, although very helpful and convenient... To Franz Wald I owe further the idea that the conception of a phase is a far more general one than that of a substance, and that the deduction of the idea of a substance, and, further, the deduction of the laws governing the nature of substances, must start from the conception of the phase. (Ostwald 1904, pp. 517–518)

The following passage from a letter to Ostwald of 16 May of the same year regarding the writing and publication of the Faraday lecture also evidences the lesser-known Bohemian scientist's priority in epistemic history.

> When you visited me last year, I did not think to take up your questions immediately and my remark about joint work was related more to the general direction. The endless discussions in Zeitschrift [für Physikalische Chemie] about the phase rule gave me cause to reflect about this, and this resulted in [...] finally steering inexorably into your wake with the result that I may well have sailed a bit faster than you. However, I do not know if you agree with my point of view. (Pinkava 1987, p. 116)

And finally in his autobiography, *Lebenslinien*, Ostwald again states with equal clarity:

> Wald had concerned himself with similar fundamental questions of chemistry, in particular the concept of pure substance, and from his position approached the same goal without being able to reach it. And once it was reached, it became apparent how astoundingly simple the matter is. (Ostwald 1927, pp. 371–372)

Principles represents a further development of the Faraday lecture,[7] which – along with Wald – is explicitly mentioned in the preface. Yet Ostwald does not simply take up Wald's ideas and develop them further. Whereas Wald almost exclusively addresses stoichiometric problems with his *analytical*, quasi *a priori* theoretical chemistry, the work in question here has remained to this day the broadest, most comprehensive attempt at formulating a positivist-operationalist theory, that is one with *minimal metaphysics*, of the *entire* range of contemporary chemistry. Over a period of several years Ostwald has with logical consistency moved

7 This is evidenced by his again taking up specific themes up to and including the examples. Alan Rocke expresses this similarly: "In 1907 Ostwald expanded his Faraday Lecture into 'an introduction to all textbooks of chemistry'." (Rocke 1984, p. 332)

away from solving practical scientific problems in chemistry with physical methods and more toward working on methodological and philosophical problems of the sciences, in particular chemistry. Thus with a certain logic he outgrows practical chemistry. This logic is reflected in his series of major chemistry textbooks: the *"Big Ostwald,"* the two-volume *Lehrbuch der allgemeinen Chemie* of 1885–1887, is still a classic textbook closely attuned to chemical reality; the *"Little Ostwald,"* that is *Grundriss der allgemeinen Chemie* (1889), has the distinctive peculiarity of containing no mathematical formulas, and *Principles* of 1907 provides a doctrine of the methodology of the philosophy of science without direct reference to the concrete realm of experience. Seen from the perspective of historical development, Ostwald writes *Principles* in the course of the logical further development of his far-reaching philosophical interests, inspired or accelerated by the works of František Wald.[8]

The reasons the author himself gives for writing *Principles* are as follows:

> Aside from the task of enriching the outer realm with new facts and experiences, every science has the task, no less important if not as obvious, of placing the known facts in relation to one another and in order. (Ostwald 1907, p. III)

At least the first part of this sentence is a description of what could be regarded as *normal science* in Thomas Kuhn's philosophy of science. However, the present approach will consider the fact that Ostwald's directive of *placing ... in relation to one another and in order* means more than merely sorting and systematizing. Rather his natural *scientist* must also be a natural *philosopher* and by conceptual analysis find and formulate generally applicable higher-order relationships beyond routine daily work. He regards mathematics, geometry, and mechanics as models or guides for proper methodological procedure in chemistry as well because these disciplines are more mature and clarified their respective fundamental relationships earlier (Ostwald 1907, p. IV). Similarly to Kant, Ostwald imagines the enormous theoretical consistency and a certain completeness that characterize the disciplines mentioned as the goals of chemistry at this point.[9] The agenda for *Principles* consequently is:

---

8 Wald himself came to the attention of officials due to the public praise in the Faraday lecture and as a result of this and further recommendation was later selected to chair the Department of Metallurgy and Theoretical Chemistry in Prague.

9 The influence of Kant's epistemology and philosophy of science on subsequent natural scientists at least in Germany should not be underestimated. With respect to chemistry, this goes beyond the achievements of Jeremias Benjamin Richter, recognized only later as the creator of *mathematized* stoichiometry (see Ruthenberg 1996b), whom Wald and Ostwald explicitly mention and who stud-

> The present work has the purpose of representing the actual fundamentals of chemistry as devoid as possible of additions that do not belong to the matter in their significance and in their relationship. (Ostwald 1907, p. V)

The methodological and phenomenological approach for placing chemistry on a rational footing is also significant for the didactics of this discipline. Ostwald was certainly aware of the *didactic dilemma* which I mentioned in the introduction: On the one hand students confronted with an area of science for the first time cannot really be familiarized with it by merely being fed definitions, principles, and laws while on the other hand these students are helpless in the presence of chemical experiences when explanations of the relationships are lacking. After the acquisition of initial illustrative material, Ostwald says, it is important to obtain an overview of the broad relationships that is free of all incidental information. He states:

> The questions of principles are the first things the student encounters at every level, and the often unconscious grasp of them from that point on determines the intellectual characteristic of the developed chemist. (Ostwald 1907, p. VI)

Thus Ostwald also hopes to influence the molding of future chemists' methods of thinking and working, and this is reflected in his choice of the subtitle *An Introduction to All Chemistry Textbooks.* In the following sections I will trace several of his specific lines of inquiry in detail, beginning with his non-atomism.

4.1.2 Why without Atoms?

In *Principles* Ostwald addresses the concept of *energy* in the very first chapter *Bodies, Substances, and Properties.* He notes that *all characteristics of the bodies are defined as their energy activities* Ostwald 1907, p. 10). Accordingly, all substance characteristics, i. e., chemical characteristics, are without exception energy phenomena and not *secondary* features inherent in substances. According to Ostwald, there is even a specific *chemical* form of energy, and he classifies the system of the sciences according to the respective types of specific energy:

> Accordingly, one distinguishes mechanics, thermodynamics, electricity, magnetism, optics (which in recent times appears to have developed into a branch of electricity), and chemistry. (Ostwald 1907, p. 12)

---

ied in Königsberg. Mach and Ostwald also repeatedly refer to the influence of Kant. Regarding relevant topics in connection with Kant, see van Brakel (2000, 2006) and Mahootian (2020).

There are also mixed forms of the types of energy mentioned with corresponding or related fields of science, and (as hinted at in the parenthetical phrase in the quotation) advances in knowledge cannot only produce mixed forms but *reductive* developments of the pure forms can also result. The situation for chemistry, from this perspective still favorable as regards its scientific autonomy, would have worsened significantly from a modern perspective: Of the fundamental interactions that modern physics distinguishes and which can be regarded as energy phenomena, *gravitation* and *weak* and *strong* interaction do not come into consideration in chemistry, but only *electrostatic* energy. From the perspective of modern *physics*, chemistry is a part of the theory of electricity.

Wilhelm Ostwald's philosophical attitude can be characterized by three basic premises: *positivism, operationalism, and energetics.* His rejection of hypothetical descriptions and explanations (disregarding energetics for the moment) combines logically with his demand to secure the determination and clarification of fundamental concepts of his own scientific field by specifying experimental procedures, that is by operationalism. The energetics dogma clearly does not align with these two ways of thinking. To appropriately evaluate the results of Ostwald's efforts, one would do well to identify those parts that are *unrelated* to this dogma.

If matter is to be ontologically subordinate and hypotheses are not to be used for the formulation of scientific concepts, then it follows logically that the atomic hypothesis – and the assumption of atomism was no more than this at least when Ostwald first began his work in this field – must be dispensed with. Beyond this theoretical aspect, Ostwald sees a further higher-order argument that makes his motivation for anti-atomism clear and which he describes in *Lebenslinien:*

> In particular, I came to realize how disproportionately more fruitful energetics was than the kinetic-atomistic doctrine which at the time had become nearly sterile and which after a short glorious ascent in the 1860s had entered the thicket of mathematical difficulties that stripped it of its freedom of movement and almost completely held back its supporters from the pursuit of new experimental paths. (Ostwald 1927, p. 178)

It is thus the knowledge-guiding *heuristic* force of non-atomistic or anti-atomistic energetics that he additionally postulates.[10] In *Principles* one may therefore look for statements in which *atoms* or *molecules* occur, but one will search in vain for statements *about* these concepts.

---

10 Paul Needham has convincingly shown that adequate atomistic explanations of chemical phenomena did not arise before the 20th century (Needham 2004). For a discussion of earlier disputes of atom skeptics cf. Buchdahl (1959), Nye (1981), Rocke (1984).

4.1.3 What Are Substances?

According to the *classifying* definition given at the beginning of *Principles*, chemistry is an inorganic natural science that concerns itself with the study of inanimate objects (*bodies*) of the external world (Ostwald 1907, p. 1). Interestingly, the author ignores the areas of contemporary chemistry that concern themselves with (parts of) living objects. In the same passage the author describes *bodies* as "... *areas of space that in a certain way behave differently than their surroundings*." This typical behavior, according to Ostwald, is the result of certain characteristics accessible to everyday experience, specifically *color*, *luster*, *form*, and *weight*. He states:

> Thus everyday experience shows me the spatial *cohesion of certain characteristics* and the sum of such experiences is brought together in the concept of the *body* [...]. It is thus a natural law that certain characteristics cohere in such a manner that they cannot be moved from place to place independently of one another but invariably move simultaneously. (Ostwald 1907, p. 2)

The fact of experience, proven by repeated observation, of the regular coexistence of certain characteristics of objects, indeed the *existence* of the bodies, is what Ostwald calls a *natural law*. However, he is aware of the provisional validity of such laws of nature. He sees them more as *guides* for explaining familiar facts or for realizing intended events, and continues:

> A natural law accordingly presents itself as the expectation of a relationship between possible experiences; this expectation is based on the fact that in all previously observed cases, this relationship was revealed. (Ostwald 1907, p. 4)

Falsification of these natural laws appears to be possible in principle but very improbable. The following is obvious: The closer a natural law like the *body law* (my term) is to everyday experience, the less the undesirable hypothesizing or the greater the desirable ontological economy (always disregarding the energetics dogma).

Ostwald distinguishes in the classic empirical sense *intrinsic* and *random* characteristics and thus obtains a further opportunity to characterize chemistry and differentiate it from physics:

> ... chemistry concerns itself with the intrinsic or specific characteristics whereas the random characteristics are assigned to physics. (Ostwald 1907, p. 6)

Examples of random characteristics he lists in the same passage include change in temperature, illumination with colored light and magnetization, quantity, and external shape. A problem with Ostwald's argumentation arises when compared with the statement made later in *Principles* that the sciences correspond to their own forms of energy (Ostwald 1907, p. 12). This poses the question as to where exactly one should seek the *chemical form of energy* and which characteristics it should have if thermal values like those crucial to chemical thermodynamics are assigned to physics. Values such as bond energy, dissociation energy, hydration energy, and enthalpy of combustion would be physical characteristics from this perspective. The entire research program of physical chemistry would be problematic for the autonomy of chemistry in this regard as it uses the methods of physics to study chemical characteristics. These methods of physics also pertain to *random* characteristics and it remains unclear how the intrinsic characteristics are depicted in this program. Even the statements made in his *Grundriss der Naturphilosophie* are hardly more convincing: Here Ostwald represents the chemical elements as independent types of energy (Ostwald 1908, pp. 169–170).[11] The *chemical form of energy* postulated by Ostwald thus remains mysterious even in his time.

For the example of silver, Ostwald specifies a group of characteristics that cannot be taken individually or altered: the solid, typically metallic state; high electrical and thermal conductivity; good chemical stability; solubility in nitric acid. Then he makes the transition to the concept of *substance*, which he derives from that of the *body* (cf. the citation and discussion in the introduction):

> Bodies that are observed [without regard to quantity and form] with respect to their specific characteristics are known as *substances*. It is thus the substances that form the object of chemistry. *Those substances that have matching specific characteristics are said to be identical.* (Ostwald 1907, p. 7)

With the last sentence Ostwald hints at the so-called *substance law* that he will later introduce. He regards this, too, as a natural law (indeed one that to this day is indirectly accepted as valid but is not commonly mentioned explicitly, let alone by the name Ostwald suggested). The author defines this law in the third chapter *Mixtures, Solutions, and Pure Substances* as follows:

> If several specific properties are the same in two bodies, all other specific properties will also be the same. (Ostwald 1907, pp. 74–75)

11 The passage mentioned reads as follows (original author's emphasis): "Whereas there is for example only one type of heat or motion energy, and in electricity only the two polar opposite forms, there are in chemistry after the greatest possible theoretical reduction at least 80 types, namely as many as there are *chemical elements.*" For the autonomy of chemistry, Ostwald mentions in the same passage its technical and therefore societal significance (an argument that again occurs among dialectical-materialist authors as a justification for the *philosophical suitability* of chemistry; see for example Simon et al. 1982).

An operational directive for identifying substances can be derived from the substance law. If there is agreement as to which characteristics are to be treated as specific and if in two specimens of substances at least two – that results from the word several – of these characteristics are the same, then the samples belong to the same species. In natural science training and in chemical research as well, this operational directive was applied then and is applied now canonically (determination of the melting or boiling point, refraction index, spectrometric measurements, etc.).

As has been noted, Ostwald fully understands the provisional, inductive character of the results of chemical research. Because there is always the possibility of discovering new characteristics and *new* substances with *new* characteristics can also be created at any time (Ostwald 1907, p. 76), chemistry is never finished. Yet he sees in the recognition of the substance law the opportunity for economizing research simply because not *every* characteristic of a substance under examination need be detected in order to characterize or identify the substance. Of course when characterizing or identifying a new substance, one cannot resort to comparing it to existing samples of the substance. Application of the substance law can thus at best lead to a relational balance. Complete acquisition of all the substance characteristics is not possible, as is the case in all of chemistry.

But what happens when one encounters an initially unknown substance in the laboratory – and this is the normal case – and is unsure whether it is present in pure form? If this substance is, say, a liquid composed of two mixable liquid components, then the characteristics, such as the boiling point, will change in a certain range depending on the composition. If the boiling point as a specific characteristic is specified or known, then, applying the substance law, the composition of the specimen of substance in question must remain unchanged in the desired range. If the sample is a solution with two components, then the substance law cannot immediately be applied (Ostwald 1907, p. 83); suitable separation processes must first be applied to ensure the presence of pure substances. Ostwald consequently examines the mixtures and separation processes more closely in the third chapter. Substance species can assume various aggregate states (in *Principles: types of forms*). Ostwald evaluates the change of a species from one state to another as a *chemical* process:

> According to our general determination, we must regard such a transition as a *chemical* process because in it a substance, the gas, with its characteristics disappears and another substance, the liquid, appears with different characteristics. (Ostwald 1907, p. 95)

As a gas condenses, new specific characteristics appear – Ostwald mentions surface tension as an example – and therefore the liquid species is strictly speaking a different substance than the gaseous one. In this manner, ice, liquid water, and water vapor must be regarded as different substances. When the transition to another state occurs at a constant temperature (melting point, boiling point, phase transformation points) and the composition of the two substances does not change, then as a rule these represent two manifestations of the same species, that is to say a pure substance (98). Here another natural law applies, which Ostwald calls the *law of continuity* and states (as previously in *Principles*) that frequent use has been made of it without it being specifically named:

> The law of continuity expresses the fact *that those properties of an object or process that are mutually dependent are simultaneously continuous or discontinuous.* (Ostwald 1907, p. 275)

For the example of water, this law means that when it freezes both the mechanical characteristics and the volume, light refraction, heat capacity, and other specific characteristics *suddenly* take on other values or that these values – like solidity and hardness – are only now definable or detectable (Ostwald 1907, p. 276).

4.1.4 Substances and Phases

In the Faraday lecture Ostwald refers to the phase concept as *far more general than that of a substance.* The concept of *phase* is introduced in the fourth chapter of *Principles: Transformation of the Types of Forms and Equilibriums.* Ostwald describes the compression and cooling of a gas, which leads initially to condensation and then further to solidification. All possible substance variants of this species will occur at some time as a result of this operation. Finally, a homogeneous (*uniform*) gas will become a homogeneous solid. He continues:

> The components of the mixture that occur as a result of such transformations are called phases. Phases are thus the uniform substances that occur in mixtures. They can be pure substances or also solutions. (Ostwald 1907, p. 117)

This formulation is certainly not particularly convincing.[12] By *components* he naturally means the homogeneous substance variants, not individual substance components. The second sentence is also confusing, to say the least. Describing substances as phases that are components of mixtures reverses the order of the concepts phase

12 Ostwald tends to use somewhat convoluted expressions on occasion. Another example: "The word 'sublimation' is commonly applied to the transformation of a vapor into a solid body." (Ostwald 1907, p. 121) Sublimation is normally defined precisely the other way around, as the transformation from a solid state to a gaseous state.

and substance as Ostwald himself understood it in 1904. Although there is nothing wrong with the content of the third sentence, one would have hoped that precisely this book would provide a clear definition of such a crucial term.[13] In the following section Ostwald introduces Gibbs' *phase rule* for the case of a single-substance system and explains its characteristics and range of application very plausibly, providing examples. It is in this part of the text that he also introduces the concepts of *allotropy* and *metastability*, which are important in chemistry. Ostwald notes that the transformations of different manifestations (the *allotropic* forms) of a substance of constant composition into one another are analogous to a change in the aggregate state. However, these phase changes are at times inhibited or extremely slow so. Thus, we also encounter substances in conditions that are not stable for them once equilibrium has been achieved. Modern science calls such conditions *metastable*. Ostwald summarizes the formation of certain products observed to occur in the transformation of allotropic substances in the law he identified, which is now known as *Ostwald's step rule* (Ostwald 1907, pp. 142–145): Of the possible products, it is not the most thermodynamically stable that forms first but that product that lies on the next higher step on the stability ladder.[14]

Ostwald describes pure substances as borderline cases of solutions (Ostwald 1907, p. 167). In phase transitions (*changes in the type of form*), solutions are changeable whereas pure substances are not. The purer a substance becomes in a certain operation, the less effect this changeability has, and finally we obtain constant measured values in the phase transformation observed. In the chapter *Elements and Compounds* the author describes it as follows:

> By definition we call pure substances those that allow changes in the type of form, that is the formation of new phases, at constant pressure and temperature values. (Ostwald 1907, p. 260)

When in a phase transformation the composition of the newly occurring phase is the same as in the previous phase, then Ostwald, as mentioned above, refers to this transformation as *hylotropic* (Ostwald 1907, p. 259). Hylotropism can occur with pure substances and with solutions. Its *region of stability* is limited in the case of pure substances yet extends over a broader range than with hylotropic solu-

---

13 Wald in his main book *Chemie fasí* is more careful with the phase concept: "[...] with the word phase [we refer to] each part of such a solution in equilibrium that is detectable with the senses and which represents a homogeneous part that is distinguishable from the others." (Wald 2004, p. 112) Schummer (1996, pp. 175–180) provides an in-depth, critical discussion of the operational applicability of the phase concept.

14 A widely used university-level textbook refers to this law (in the description of the polymorphism of sulfur) under the name *Ostwald's step rule* (Holleman & Wiberg 1964, p. 190).

tions, which exhibit this characteristic only at individual points (azeotropes, eutectics).[15] Note here that the obviously enlightening concept of *hylotropism* has not been accepted by the community of research chemists. If one goes beyond the respective energetic region of stability of pure substances by applying more or less extreme conditions, several other pure substances can occur. Ostwald calls these processes *chemical processes in the narrower sense* (Ostwald 1907, p. 263). If one applies this process repeatedly, one will obtain substances that cannot be further separated into mixtures or solutions using the given methods: the *elements.* As they are introduced, Ostwald also calls them "... *simple,* or better *undecomposed* substances" (Ostwald 1907, p. 265).

> An element is a substance that cannot be transformed into another non-hylotropic substance within the entire region of attainable energy influences. (Ostwald 1907, p. 267)

Whereas in most contemporary (and also modern) representations, chemistry is described from the end and elements are treated as preexisting and complete, Ostwald, too, emphasizes in *Principles* explicitly and in methodological-operationalist terms that they must be prepared.

4.1.5 Reactivity

Conventionally, chemical reactions are usually described in static terms. By "static" I mean that initially little attention is paid to the course, the *temporality* of the processes, which is closely related to the mutability of the substances. This has to do with the strong, centuries-long current in intellectual history associated with mathematization and physicalization in which *form* is given higher priority than *substance,*[16] but it is also related to the seemingly necessary systematic subdivision of the field of knowledge into smaller, more tractable, and more readily describable areas. This process of subdivision has generally overemphasized structural considerations. Yet if one regards chemistry as a science *essentially* interested in *temporal* processes, namely substance metamorphoses, then it can be expected that any fundamental theoretical chemistry, and naturally the philosophy of chemistry as well, must address the question of *reactivity.* This question is of interest in the present context because on the one hand Ostwald, as a proponent of energetics, first and foremost focuses on the *potentiality* of natural processes, and

---

15 In the discussion of *chemical species* Paul Needham provides a readily comprehensible outline of this part of Ostwald's theories (Needham 2002, pp. 214–218).

16 *Dematerialization* has already been discussed at length, cf. Schummer 1996.

this and *temporality* represent two crucial perspectives or approaches (see Chapter 6). On the other hand, seen from the early phase of *Principles*, he was later awarded the Nobel Prize in chemistry and this in recognition of his achievements in *catalysis*, a field generally considered as belonging to chemical kinetics.

In the description of the reversibility of chemical reactions Ostwald notes the asymmetry that fundamentally characterizes the chemical realm: Breaking down compound substances, that is analysis in the strict sense, is easier to do than synthesis. This is the reason why the earlier name for chemistry in German was *Scheidekunst* (and to this day is *scheikunde* in Dutch) which essentially means the art of separating substances. Here Ostwald offers no explanation for this asymmetry, neither anything that might be plausible today nor a thermochemical explanation of his own using entropy arguments. Such an argument would go like this: It is easier to destroy substance structures than to create them because there are more ways to achieve the goal of breaking a substance down than to synthesize it.

The standpoint Ostwald offers in *Principles* is not very clear regarding the *chemical* law of the conservation of elements. On the one hand, he admits that the elements in the substances *potentially* remain unchanged in chemical operations and that elementary analyses would lead to undefined results if the law of the conservation of elements did not apply (Ostwald 1907, p. 272). On the other hand, further below in the discussion of chemical formulas (that is empirical formulas), he states:

> As the symbols of the elements compose the symbols of the compounds, they automatically suggest the notion that the elements are still physically present in the compounds as the symbols are in the symbol of the compound. On the other hand, it is the distinctive feature of chemical processes that in them substances disappear and others with other characteristics emerge [...]. The expression that the characteristics have disappeared yet the elements are nonetheless still present in their "nature" is too vague to be regarded as scientific. (Ostwald 1907, p. 395)

The change of the specific characteristics and disappearance of the characteristics of the elementary components in the formation of a chemical compound leads Ostwald (like Mach) to dispute the continued existence of the elements in a compound.[17] Even the concessions of the conservation of mass, which enables stoichiometric calculations, and the reversibility of chemical reactions discussed previously do not convince him. Concessions of this sort would lead Ostwald perilously close to atom-

17 Aristotle denied the further existence of the elements in compounds. In an essay about the discussion of *chemical species* Paul Needham, in his analysis of the *Faraday lecture*, also comes across this point with respect to the comparison of Ostwald and Aristotle (Needham 2002, p. 217). I feel Ostwald's statement on this topic in *Principles* is clearer.

ism. He prefers a different analogy: Amounts of money deposited in a bank can be withdrawn at any time. Yet what has happened to the deposited money in the meantime and whether one will receive the same coins and banknotes again is uncertain (Ostwald 1907, p. 396). Although his contradictory position resulting from his fundamentally dogmatic attitude is evident at this point, an important aspect of his standpoint still applies in modern chemistry: Here, too, certain characteristics of the elements are assumed to change when they form a new substance. For example, the carbon atoms in an ethane molecule are assigned a different electronegativity and a different oxidation number than those in an ethyne molecule. Thus one can speak of the strict *identity* of an element in a compound from the *physical* perspective but not from the *chemical* perspective.

In the ninth chapter, *Rate of Reaction and Equilibrium*, Ostwald explicitly addresses the temporality of the chemical process. Here he provides an explanation for the difference between and the different treatment of *potentiality* and *temporality:*

> Changes in energies present in the form of a reduction in their differences are in the final analysis what everything that happens leads to or what is so inextricably linked to everything that happens that it must be regarded as the essence of the process. (Ostwald 1907, pp. 450–451)

Because the entities strive toward temporally invariant resultant states, the scientific descriptions of these states, insofar as these descriptions refer to their result, are essentially *timeless*. According to Ostwald, the state of a substance system in equilibrium is determined exclusively by energy, because the time in which the equilibrium is achieved is indefinite (Ostwald 1907, p. 467). Rates of reaction can even be controlled over broad ranges if *catalysts* are used. In a passage that for Ostwald is unusually speculative with the subtitle *Ideal Catalysts*, he examines the possibility of the presence of a *universally catalytic* effect in nature. That is to say "that all reactions that actually occur would be caused by the presence of minimal quantities of catalytically active foreign substances" (Ostwald 1907, p. 464). However, the word *caused* warrants critical restraint. If energetics has priority and the catalyst does not change the equilibrium situation of the process, then one cannot speak of *causation* but more properly of *activation*. However, Ostwald rejects this term although that contradicts his fundamental attitude (see Chapter 6).

Ostwald describes how chemical equilibrium is achieved via reaction rates; he uses the so-called *kinetic* approach to introduce the *law of mass action* (Ostwald 1907, p. 475). This law is based on the equivalence of the competing reaction rates of the reaction and reverse reaction in equilibrium. Before equilibrium is reached the reactions can proceed at different rates.

> With this multiplicity of reaction rates, which can vary between zero and infinity, a far greater multiplicity than our previous observations could predict emerges for the chemical entities as well. States that are not at all equilibrium states yet which change with infinitesimal reaction rates can for this reason appear to us as equilibrium states. (Ostwald 1907, p. 468)

The phrase *between zero and infinity* should not be taken literally and is better read as *varying over large ranges.* The phenomenon described in the last sentence of this quotation is the *metastability* mentioned previously. Ostwald himself does not use this term at this point. Interestingly, in *Principles,* he dispenses with the *energetic* derivation of the law of mass action (via comparison of the chemical potentials; cf. Planck's approach in Chapter 6), which in contrast to the kinetic derivation is generally valid.

4.2 Reactions of the Scientific Community

Because he is very well known and extremely well integrated into the scientific network of his day, Ostwald's worldview and his particular perspective on chemistry naturally provoke reactions. Only a few representative examples will be named here insofar as they pertain to the chemistry-related content, mentioned above, of *Principles* or its precursor, the *Faraday lecture.*

Svante Arrhenius (1859–1927) attacked Wald's and Ostwald's claims.[18] He criticizes Ostwald for his attempt to derive the law of conservation of mass and the law of definite proportions as follows:[19] Given three elements A, B, and C that form binary compounds AB, BC, and AC) and a ternary compound (ABC). If we form ABC from AB+C or AC+B, then according to Ostwald it follows that both products result in the same compounds. Thus he concludes that with the assumption of the concept of the chemical individual (elements and compounds) that can be traced back to the empirical law of the conservation of the elements, the laws of conservation of mass and of definite and multiple proportions necessarily follow. Arrhenius contests the deductive character of these derivations but deviates from Ostwald's text right at the beginning of his discussion:

18 See Arrhenius (1906, pp. 34–41).
19 See Arrhenius (1906, p. 39). Cf. in this regard Ostwald (1904, pp. 518–519).

> Ostwald, too [like Wald], seeks to derive the law of multiple proportions and of constant compound weights in various compounds from the law of definite proportions. (Arrhenius 1906, p. 39)

The Swedish chemist also maintains that the proportion of B in the formation of the product ABC does not necessarily have to be the same as in the starting material AB and concludes that the resulting simple numerical proportions are *facts of experience.* In Ostwald's favor one can note that he maintains that the stoichiometric laws result from the empirical notions of conservation of the chemical species in certain regions of stability. Thus Ostwald does not deny the experiential character of these derivations, merely the point *where* reference is made to experience. Moreover, Ostwald does not deny that other compositional proportions can occur in the formation of compounds (this is precisely the basis of the notions of multiple proportions); he only says that *if* the same substance (ABC) results in two different ways (AB+C and AC+B), *then* the proportions in this substance are the same. The argumentation that Arrhenius presents here is not very convincing and appears to be more of an *ad hoc* attempt. However, it is interesting to note that he does not fall back on the atomic hypothesis in his *specific* criticism of the deductive character of the stoichiometric laws under discussion here. With respect to these laws, he only says of Ostwald's underlying *energetic* hypothesis in his introduction:

> But the great difficulty of working with this hypothesis stands in stark contrast to the simplicity that characterizes deductions from atomic theory. Yet because a theory like a tool is better the easier it is to work with, the notion that matter consists of swirls of ether has not been widely accepted by chemists. As far as I know, this idea has not led to the discovery of any new phenomenon or any new law [...]. I am therefore of the opinion that it is more practical to retain Dalton's simple atomic theory than to replace it with the rather academic hypothesis of swirls of ether. (Arrhenius 1906, p. 34)

Arrhenius takes a pragmatic standpoint and in a certain sense turns the tables on the attack against atomism by, on the one hand, noting the advantages of the atomic hypothesis in terms of theoretical economy and on the other hand pointing out the clumsiness and lack of productiveness of the energetics hypothesis. This dispute about two different metaphysical approaches notably did not affect the lifelong friendship between the two great opponents.

As a young scientist at Vienna's Institute of Radium Research, the Austrian Friedrich Adolf Paneth (1887–1958), a pioneer of radiochemistry, publishes a systematic overview of the element and atom concept in the *Zeitschrift für Physikalische Chemie*, edited by Ostwald. At least at one point in this essay he also critically examines Ostwald's *substance law* (Paneth 1916). He explicitly refutes the validity of what Ostwald termed a natural law – Paneth calls it a *rule* – on the grounds that *isotopy* (which Ostwald does not mention in *Principles*) has in the strict sense rendered it invalid. However, Paneth then concedes that a strict interpretation – *all* characteristics of two samples of substances in question are identical – "would lead to a complete debasement of the element concept for chemistry" (Paneth 1916, p. 182). He therefore suggests requiring the identity of the *chemical* characteristics and provides nothing less than a classic definition of the element concept:

> Two elements are designated with the same name if they cannot be separated by any chemical process once mixed with each other. (Paneth 1916, p. 183)

It should be noted here that this definition is independent of the atom concept; it is more of an *operationalist* or *phenomenological* definition, but one that is generally accepted and thus in any case compatible with Ostwald's definition quoted earlier (Ostwald 1907, p. 267). Despite his knowledge of (isotopic) pure elements and mixed elements, the atomist and radiochemist Paneth feels such a definition, although old-fashioned even at the time, is perfectly adequate. Yet in a lecture by Paneth in Königsberg 16 years later there is no longer any consensus. In the printed version of the lecture he states in a footnote:

> Wald's and Ostwald's observations, which move along a line deviating from the usual notion of element and following their own anti-atomistic thinking, shall be completely disregarded here [in the context of the discussion of old atomism]. (Paneth 1931, p. 108)

At this time, as far as Paneth is concerned, there is no compatibility between the standpoints and presumably also no patience (*deviating from the usual notion of element*) with the anti-atomists on the part of physicalist mainstream chemistry; this can only mean that atomic physics' definition of elements as types of atoms with the same atomic number has already become generally accepted.[20]

Another successful physical chemist, the Belgian Jean Timmermans (1882–1971), later had a more favorable but nonetheless critical attitude toward the thermodynamic phase-centric concept of chemistry. He quotes the Faraday lecture and Wald's writings of 1897 and, referring directly to Gibbs' phase rule, states:

> This rule is based on thermodynamics; the results to which it leads acquire much greater significance from their certainty and the extent of their field of application. The persevering efforts of Wald are largely responsible for the acceptance of this viewpoint, and it definitely acquired the right to be included in the exact sciences after its masterly discussion by Ostwald in his Faraday Lecture of 1904. (Timmermans 1940, p. 12)

20 I will discuss Paneth's own position in detail in Chapter 9.

He is certainly overly enthusiastic in portraying the reception of Wald's conception in such positive terms. In any case, this example demonstrates, first, that the thermodynamic approach to a chemical substance theory has indeed been received and, second, that this reception can also take on the form of theory pluralism. Timmermans initially presents the concept and points out its limits yet forgoes any attempt to replace the thermodynamic concept with the atomist one (as Paneth does later). He initially describes the limits of the concept in light of the *optical activity* of enantiomers (optical isomers):[21] Rotating the oscillation plane of polarized light in one direction or in the opposite direction by means of two such enantiomers cannot be represented thermodynamically (nor quantum mechanically, see further below). On the other hand, once one has available a method of non-thermal separation of a mixture of two optical isomers, the so-called racemate, then Ostwald's statements about determining the different identities of two samples of substances apply analogously. Successive mixing of the two samples of substances constantly changes the species-specific characteristic that is to be measured, namely optical activity. Other methods of specifying enantiomers result from their biochemical activity, which usually differs. In addition to the enantiomerism mentioned here, Timmermans notes that the problem areas of application of the operational-thermodynamic method of characterizing substances include the so-called *bertholides*, intermetallic phases for which no clearly defined stoichiometric formula can be specified.[22]

4.3 Conclusions

In his work in the philosophy of science, *Principles of Chemistry,* Ostwald seeks to place chemistry on a phenomenological-positivist footing that is significantly broader than that of his erstwhile companion Wald. Ostwald's theory of the sciences is not devoid of metaphysics although he himself does not see it that way. His

21 I cannot agree with John Bradley's opinion that Ostwald's program would fail with (normal) isomerism (Bradley 1955, p. 42). Ostwald is indeed venturing into dangerous territory by using these fictitious valence structural formulas, expressions that are foreign to his theory. Yet he explains that this is a "commonly used procedure in chemistry" and makes clear that here it is only a matter of comparing the two approaches and assessing their compatibility. He states: "One generally calls those substances isomers which have the same composition yet have different characteristics and retain this difference even with changes in the type of form." (Ostwald 1907, p. 499) Ostwald's treatment of isomerism is logically consistent; he treats it analogously to the types of forms of pure substances and to allotropy.

22 The passage mentioned in Timmermanns is also discussed in Schummer (1996) and van Brakel (2000).

expositions in *Principles* are nonetheless very illuminating because they present the entire range of fundamental chemical phenomena in a methodologically well-ordered fashion. This occurs at a time of scientific upheaval (the late 19th century) when an orientation in the philosophy of science and natural philosophy is a crucial topic to everyone. Ostwald himself must create many of the concepts applicable in or necessary for such a methodological argumentation. He formulates or describes what he calls "natural laws" such as the *substance law, law of continuity,* and *step rule,* and he introduces the concept of *hylotropism.* Although hardly any of these concepts have explicitly gained acceptance in the chemical sciences, their clarity continues to be evident and indisputable even from a modern perspective. To this day, no one in chemistry's realm of experience has ever really worked with atoms, rather with portions of substance. For this reason it is not only useful but also methodologically correct and necessary to conceptually clarify the appropriate region of phenomena. One of these clarifications is Ostwald's proof, first presented in the *Faraday lecture* and repeated in greater detail in *Principles*, that the stoichiometric laws, if one proceeds from the modern concept of the element, result *deductively*; they are thus not obtained empirically in the classic sense but have normative character. The fundamentals of chemistry are given a further important clarification by the operational-phenomenological approach to describing the concept of substance. This approach has its limits, yet compared with atomism, which is far removed from the realm of experience, it allows an introduction to the crucial fundamentals of chemistry that is less overloaded with theory.

One obvious disadvantage of Ostwald's thinking is indisputably his at times efflorescent energetics dogma. It is quite probably the case that this dogma is the reason why the plausible, so to speak *internal* parts of his theory of chemistry failed to gain greater acceptance in the scientific community. Yet upon closer examination, we find we can keep the energetics dogma out of broad areas of inquiry in the philosophy of chemistry because it has no effect on several of the crucial propositions, for example the natural laws mentioned. Besides – especially in light of thermodynamics, quantum theory, and quantum mechanics – nothing other than *energetics* has dominated theory in modern natural science since the end of the 19th century. It is only that no one calls it that anymore, and no one turns up their nose at it – quite the opposite. Energetics in this sense is still present, and the question future historians of science will have to explore is whether, for example, the doctrines of *atomism* or those of *energetics* have been more fruitful for chemistry – and whether or not we can finally sort out the differences between the two.

Fig. 10: The "brown ring test" for nitrate: An example of the systematic art of qualitative analytics by manufacturing a metastable reaction product. Left: The single-phase nitrate-containing acidified specimen with iron(II)sulfate added. Middle: The same specimen with concentrated sulfuric acid carefully added so that the acid forms a layer below the solution, that is a two-phase specimen (the brown ring of the desired product at the phase boundary indicated the presence of nitrate ions. Right: The same specimen after shaking (if one does not add the sulfuric acid carefully, then even without nitrate the solution can look exactly like this or may appear light yellow; in any case the test result would not be conclusive). See the more detailed explanation in the following text.

5 No Analysis without Synthesis

My aim in this chapter is to get to the bottom of the "analytic-synthetic ideal" in chemistry.[1] My specific question is: What exactly is the relationship between analysis and synthesis in chemistry? In the modern philosophy of chemistry the relationship between analysis and synthesis has hardly received adequate attention. The few exceptions include Schummer (1994, 1996, 2002), Laszlo (1998), and Jacob (2001). Here I will argue for the proposition that chemical analyses invariably include synthetic (manufacturing) aspects. Following a brief historical introduction to the term "analytic-synthetic ideal" used in the history of chemistry, I will go on in the second section to present the definitions the study requires and proceed to characterize analytical chemistry. The third section presents historical and modern examples, while the fourth section presents the philosophical discussion.

5.1 The Analytic-Synthetic Ideal

"Solve et coagula" is a central methodological principle of alchemy (cf. Gebelein 1991) which, greatly simplified and modernized, may be rendered as "analyze and synthesize." In Andreas Libavius' (ca. 1550–1616) textbook *Alchemia*, a work that is as practically oriented as it is didactic and theoretical, we find the following statement:

> Alchemy is the art of perfecting magisteries and extracting pure essences from mixtures by separating the corpus. (Libavius 1597, p. 1)

The perfection of "magisteries" refers to the myriad processes for manufacturing or "improving" certain characteristics of substances. Whereas Libavius refers more to synthetic aspects in the first part of this statement, the second part concerns analytical considerations. Libavius expresses this interesting idea about the nature of chemical substances:

> A chemical species is that which is perfected in the operations of alchemy presented in the doctrine of handicraft processes. (Libavius 1597, p. 119)

It is thus the actions of the experimenters that bring the substances to the desired quality. Alchemy understood in this manner is thus in large measure a systematic handicraft. A good hundred and thirty years later Georg Ernst Stahl (1660–1734),

1 I use here an expression introduced by Mimi Kim (2014).

https://doi.org/10.1515/9783111401041-008

the originator of the phlogiston theory, defined the science of substances in a manner quite similar in principle to Libavius:

> Chemistry, also known as alchemy and spagyrics, is an art of separating the mixed or composite or aggregated bodies into their principia, or recomposing them from such principia. (Stahl 1729, p. 1)

Here, too, mention is made of an interplay between separating and composing. A general concept of substance such as can be applied today did not yet exist at this time (gases were only included later). Very much like Stahl, other early 18th-century chemists articulated the analytic-synthetic ideal of their age (which has its precursor in the motto of alchemy "solve et coagula"), for example, Etienne-Francois Geoffroy (1672–1731):

> One is never sure, in fact, of having decomposed a Mixt into its true principles until one can recompose it from the same principles. This reestablishment is not always possible. When it is not possible, it does not necessarily work against the analysis of the mixt, but when it is successful, the analysis is demonstrated. (Geoffroy 1704; quoted according to Kim 2014, p. 123)

Although we should refrain from automatically translating "mixt" as "compound" and "principium" or "principle" as "component," the methodology advocated here becomes clear: Composite chemical individuals are separated into their component parts, and their existence is verified where possible by *resynthesis.* Whereas this method with its analytic-synthetic conception was practicable in certain experimental settings,[2] it remained a desideratum for the principles, especially phlogiston. To this day the pure principles have (naturally) remained inaccessible; even the purest form of water is not identical to the "Aristotelian" element of the same name. An important methodological consideration here is Geoffroy's statement that unsuccessful resynthesis (recomposition) does not necessarily nullify the success of the experiment. Thus the chemical analytics of the time included synthetic operations (here restoration). The verification by synthesis was later propagated by Berthelot under the rubric "total synthesis" and remained until well into the second half of the 19th century the dominant method of structural clarification, i. e., identification, of large complex molecules of natural substances.[3]

---

2 An example from Stahl's wealth of experience is the breakdown of mercuric sulfide into mercury and sulfur or, respectively, the synthetic recovery of the former from the latter (cf. Strube 1984, pp. 46–47).

3 The following was said of Robert Burns Woodward (1917–1979), a modern master of total synthesis (Nobel Prize 1965): "The syntheses of quinine, cholesterol, cortisone, strychnine, reserpine, chlorophyll, and more were then perceived as tours de force in which knowledge of reagents, as-

Against the background indicated here it has become common among the experts in the history of chemistry to refer to an *analytic-synthetic ideal* of chemical substances (cf. Kim 2014, Klein & Lefèvre 2007).[4]

5.2 Definitions

5.2.1 Synthesis and Analysis

Whereas in the conventional chemistry-related understanding of the terms, "synthesis" as "composing" or "manufacturing" is given a relatively uniform meaning,[5] "analysis" has at least two, namely first the more general meaning of a "study" and second the specific one of "breaking down."[6] Both characterizations find their philosophical correlates as *interpretative* and *decompositional* variants, respectively (Beaney 2021). It is important to differentiate the practical (experimental) and theoretical (conceptual) forms for the field of chemistry. Claus Jacob (2001) makes the helpful suggestion of an expanded notation. An example of a theoretical analysis – Jacob refers to it as analysis$_1$–would be the stoichiometric interpretation of a usual

tuteness, ingenuity, and imagination allowed the art of the synthesizing chemist to rival the synthetic mastery of life." (Bensaude-Vincent & Stengers 1996, p. 254).

4 See also Schummer (1996, p. 127; original author's emphasis), with a direct reference to Stahl: "It is not only interesting that Stahl unmistakably interprets the principles as substances that can be experimentally isolated but also that he depicts the entire variety of substances based on the elements in an operational, hierarchically structured framework that must invariably rest on the two fundamental operations, *analysis* and *synthesis*; this twofold operational requirement for the identification of the substances remains largely binding until the mid-20th century." Also referring to Stahl, Elisabeth Ströker (1982, p. 89; original author's emphasis) states: "If it is therefore, in later parlance, 'analytical' and 'synthetic' lines of inquiry which principally characterize the chemical science, then as *Chymia experimentalis* it must primarily provide insight into which substances the bodies are composed of. However, these may only be regarded as indisputably chemically identified when it has been demonstrated that they can be reconstituted from those substances into which they can be broken down."

5 The manufacturing aspect is also occasionally interpreted in such a way that "synthetic" is equated to "artificial" (cf. Bensaude-Vincent & Simon 2008, p. 101).

6 In the history of chemistry, Robert Boyle (1627–1691) is usually mentioned as the first to have used the term "analysis" (cf. Szabadvary 1966). The concept pair analysis and synthesis (or better analytic and synthetic) has a past to contend with in the history of philosophy (cf. Tugendhat & Wolf 1983). However, that discourse is not the topic of this chapter although the comparison with Quine's *Two Dogmas* appears tempting. In the following discussion, instead of referring to "analytical chemistry," that is chemical analysis as an examination, I will use the now common expression "analytics."

chemical reaction equation, that is a statement about the expected proportions of the components involved. Take for example the formation of water from the elements:

$$H_2 + ½\,O_2 = H_2O$$

Jacob regards as analytical$_1$ the information that 1 mol of hydrogen and ½ mol of oxygen produce 1 mol of water, assuming a complete reaction. Accordingly, this excludes the fact that significant deviations from these quantitative relationships occur in reality. Read from left to right, the process would also be a synthesis$_1$, and read the other way an analysis$_1$ as regards the respective products. A theoretically constructed connection of the classification analytical$_1$ or synthetic$_1$ can be formally ("syntactically") completely acceptable without any hope of "semantic" correctness, i. e., feasibility. This means that the capacities of formal language exceed the experimental capabilities of laboratory practice. Because experimental operations, that is chemical reactions, relate to real substances, Jacob distinguishes them from the formal descriptions by referring to them as analysis$_2$ and synthesis$_2$. He regards those chemical entities that are manufactured for analytical purposes (such as the molecular fragments in mass spectrometry) as analytic$_2$ compounds. Jacob consequently offers the unusual variant of synthesis for analytical purposes. In the following section I use the term synthesis in an extended sense of "manufacture." Here I am interested not only in the conventional, intentional production of substances from more or less well-known constituents.

5.2.2 Self-Descriptions of Analytics

Among modern specialists, analytical chemistry is often presented without any reference to substance.[7] Representative of the tenor of a large part of the scientific community (being selected and awarded by a jury of specialists), the winner of a competition, a professor of analytical chemistry, maintains:

> Analytical Chemistry is the cognitive sense of every natural science discipline grasping for an unbiased chemical reality of the microcosm, whether it is an isolated sample of matter in our hands or a star far away in the universe. It is the type and number of atoms, together with their arrangement in three-dimensional space, which determine the properties of matter. (Cammann 1992, p. 812)

---

7 This correlates with a typical characteristic of modern chemistry. Janich (1994b) describes this characteristic as a preference for a "beginning from the top," that is from each of the latest models.

Such a statement moves the intrinsic, supposedly unbiased "chemical reality" into the microcosm of the non-observables and turns the existence of atoms and their countability and spatial arrangement into what amounts to a metaphysical prerequisite. The empirically necessary work is subordinated to this "chemical reality" and at best appears in such narratives as "sample preparation." In other words, the qualitative and phenomenological analysis is disregarded or simply assumed to be somehow self-evident. As was said, substances are no longer mentioned. A further variant of self-description is that of another analyst, Klaus Danzer. He states:

> The object of Analytical Chemistry is to obtain chemical information about materials concerning their qualitative and quantitative structure. (Danzer 1992, p. 827)

Danzer uses the more neutral word "materials," which is closely related to "stuffs" or "substances." This makes the connection to the world of real life more possible and plausible than do microphysical catchwords and an emphasis on theoretical entities. What exactly "chemical information" and "qualitative and quantitative structures" are remains unanswered, yet one need not necessarily understand it as associated with an extreme physicalist attitude. Danzer describes the relationships in a scheme: First, the problem is defined qualitatively and quantitatively. Then, the essential steps are the taking and preparation of a sample to obtain a test specimen for the subsequent measurement. The measured data or raw data are acquired by operations such as weighing, volumetric measurement, measurement of charge or electricity, temperature measurement, etc. (cf. the examples discussed). By applying suitable calculations (data processing), the result is obtained from these data. This result is also compared to the expected values and evaluated with respect to the definition of the problem. In the worst case, the cycle is repeated from the start allowing for the experiences with the initial cycle.

A further important empirical fact crucial to the characterization of a substance was described by Wilhelm Ostwald in his *Wissenschaftliche Grundlagen der Analytischen Chemie.*[8] Ostwald notes that when two substance specimens are compared, it is not possible to prove that they are identical in every characteristic, and continues:

---

8 As expected, Ostwald provides a very neutral and non-atomistic definition of analytics as "the art of detecting substances and their components" (1894, p. V).

> We regard such a proof as superfluous due to an unspoken induction of very great generality, for we know: *if two substances conform fully in several characteristics, they do so in all others as well.* (Ostwald 1894, p. 3; original author's emphasis)

Ostwald regards such an "induction of very great generality" as a natural law and derives a plausible operational methodology from it. In some cases, he states, even a single characteristic is sufficient to identify a substance. I would take issue with this because under given conditions at least the classification as an aggregate condition invariably involves further observations or measurements, and that is a substance characteristic. This in any case is the conclusion we can draw from the helium example (see below; the fact that it is a gas could only be deduced from comparison with an actual sample[9]). In actual laboratory practice, there is no substance about which we can know only one characteristic (given that the conditions allow sensory experiences). In any case, Ostwald feels that in most cases several different determinations of *state* or *process* characteristics as he calls them are required:[10] "The probability of error decreases rapidly with an increasing number of independent tests." (Ostwald 1894, p. 4)

5.2.3 Quality

The determination of identity by its nature belongs to the realm of qualitative analysis and suitable "independent tests" arise as the process progresses. For example, when a substance specimen is tested for flame coloration, its aggregate state and color (and often much more) are already known. If gas chromatography/mass spectrometry (GC/MS), as mentioned below, is used, then retention times and mass spectra are also known. It may be a platitude but nonetheless I feel it warrants mention again: An adequate amount of qualitative knowledge is an essential prerequisite for any quantitative determination. The role of qualitative work in analytics is accordingly a fundamental one. For example our scientific understanding of acids and bases is based on common perceptions such as taste, changes in color (see indicators in titrimetry), and formation of gas or foam (in neutralization re-

9 Not all chemical species are material in nature in that they can be presented in the form of concrete macroscopic samples.

10 An essential characteristic analogous to, say, a fingerprint in criminology is something we will not find in chemistry. In cases of examination of non-material chemical species such as certain radicals or individual molecular fragments in mass spectrometry, the selective artificial design of the experiment replaces the necessity of "independent tests."

actions).[11] All of these impressions are based on the formation of certain substances, that is to say they have a synthetic basis.

A striking feature of the qualitative investigation methods in chemistry is that it is fundamentally incorrect to reduce our understanding of chemical analytics to compositional tasks. I understand "compositionism" as that branch of research in which the composition of the specimens under study is the focus of interest.[12] It is the foundation of the theoretical and methodological shift that is often referred to as the "chemical revolution." Compositionism is the program that in chemistry experienced its zenith during the course of the 18th century with Lavoisier and had superseded "principlism," at least in the conventional interpretation. In the latter, chemical substances are described and classified according to certain phenomenological characteristics. There are no elements in the modern sense in this system. In conceptual terms, the qualitative and quantitative investigative methods here stand face to face (principles vs. composition).

It is obvious that since about 1800 compositionism has predominated in professional analytics, at least in the image it projects of itself (see Cammann's definition quoted above). Even general chemistry has shown a clear trend toward abstract microphysical interpretations in its attempts to define itself. Yet when it is a matter of the actual relationship between analysis and synthesis on the one hand and quantities and qualities on the other, then we must not overlook the fact that the characterization of chemical entities, even or particularly when quantitative methods are applied, cannot make do without qualitative and synthetic aspects however strongly one may desire an unassailable abstract mathematical description. The empirical formula H_2O alone does *not* represent a sufficient description of water.

5.2.4 Quantity

The Danish chemist Edmond Rancke Madsen (1910–1987) presents in his dissertation a systematically relevant attempt at an overview of quantifying methods. In it

11 Ruthenberg & Chang (2017) provide a detailed study of the early forms of acid characterization.

12 I take this expression from Hasok Chang (2012, pp. 37–42), who uses it to distinguish from the "principlism" of the 17th and 18th centuries. The attitude behind compositionism regards analysis primarily if not exclusively as cognitive "breaking down" and is thus one main target of my argument in this study.

he distinguishes three types of analysis: isolametry, normometry, and titrimetry (Rancke Madsen 1958, p. 209).[13]

Rancke Madsen uses the term *isolametry* to designate the variant of analysis in which the target chemical entity is isolated completely or to a known degree so that it becomes possible to "directly" determine it. He understands *normometry* as the determination of a component by observational or measurement comparison to known norms (standards) based on previously established dependencies of certain qualities of the concentration of the analyte. The observed qualities can be a) physical, b) chemical (see catalytic effects), or c) biological in nature. *Titrimetry* is then described as a determination of the reaction capacity of a chemical entity, whereby two criteria must be fulfilled: i) The stoichiometry must be known in sufficient detail (see the examples given) and ii) an indication must be ensured. Rancke Madsen notes that a titrimetric process includes two measurement operations, the quantification of the titrator and the measurement of the endpoint of the reaction, that is the indication. He states that titrimetry is therefore different from the other two basic analytic variants. Additionally, he distinguishes direct titration from the converted and indirect forms. If a titration is "converted" in this sense, another component will be determined by a corresponding reaction sequence insofar as the necessary information about the references of the titrands is available. If the surplus of another component is measured after the reaction with the titrand, then according to Rancke Madsen this would be an "indirect" measurement (in modern parlance the latter is referred to as "back titration").

Rancke Madsen does not explicitly mention the synthetic aspects required in all the operations mentioned, rather he seems to assume they are self-evident. Nevertheless, he goes on to discuss the characteristics of the analysis variants in greater detail by introducing the concepts "sensometry" and "quantimetry" in addition to the operational classification. These two methodologies correspond to the conventional distinction between qualitative and quantitative analysis, that is examination *without* the aid of instrumentation and *with* it.

With respect to sensometry, the author states that it is used only in exceptional cases in isolametry because that method allows only rough quantitative estimates at best. In this realm of analysis the gravimetric and volumetric processes predominate; at least that was the case when his study was published. Normometric processes can have a sensometric character, as is the case with colorimetric tests. Earlier processes also utilized the senses of taste, smell, hearing, and occasionally touch. However, as a rule, the normometric field is characterized by the use of

13 The author of this study, long parts of which relate to the history of chemistry, notes that his selection is not necessarily complete.

quantifying processes. I will return to the significance of sensometry further below. The concepts presented here are intended to provide an initial orientation for the discussion of additional examples.

5.3 Examples from Analytical Practice

The following examples show synthetic (manufacturing) aspects of analytics. They are listed roughly in order of increasing complexity.

5.3.1 The Brown Ring Test

Many students of chemistry (among them the author) have wondered about how the elaborate method of detecting nitrate in aqueous solutions came into being. This method was certainly known already in the 19th century:

> If one adds to a solution of a salt of nitric acid the same volume of concentrated sulfuric acid that is free of nitric acid and nitrogen peroxide, lets it cool, and then pours a concentrated solution of iron vitriol onto it in such a manner that the liquids do not mix, then the contact layer will first turn purple, later brown, or with very small quantities of nitric acid a reddish color. If one mixes the liquids, a brownish purple clear liquid results. (Fresenius 1866, p. 233)

As practically all salts of nitric acid (nitrates) are water soluble, it is not possible to use precipitation reactions for detection purposes as for some of the other more important salts. Carl Remigius Fresenius (1818–1897), from whose widely used textbook the quotation above was taken, explains the phenomenon as follows in the same passage:

> The nitric acid is broken down by the ferrous oxide; three-fifths of its oxygen joins the protoxide and transforms part of it into oxide; the remaining nitric oxide combines with the not yet more highly oxidized iron oxide salt to form a peculiar compound which dissolves in water with a brownish black color. (Fresenius 1866, p. 233)

In modern introductory chemistry lab classes the experiment looks only slightly different (cf. Figure 10): A bit of iron sulfate ($FeSO_4$, “iron protoxide”) is added to a weak nitric acid solution and concentrated sulfuric acid is carefully added so that it forms a layer below the dilute solution. When nitrate (or nitric acid) is present, a reddish brown ring occurs in the boundary layer between the dilute solution and the concentrated solution, hence the name brown ring test. Many generations of students will have wondered what exactly goes on here. The process is quite complex, and this is because of the diverse nature of nitrogen-oxygen compounds. There is an entire family of nitrogen oxides and at least one of them, nitric oxide (NO), is involved in the brown ring test. As had been discovered as early as the first half of the 19th century, this represents a redox reaction: Iron ions in oxidation state 2 (Fe^{2+}) are oxidized to oxidation state 3 (Fe^{3+}) whereas the nitrate ions are reduced, namely to those NO molecules. The excess Fe^{2+} ions then form the iron nitrosyl complex $[Fe(H_2O)_5NO]^{2+}$. This complex is only metastable, for which reason one must take care to avoid shaking the test tube if the qualitative test is to produce reliable results.

An important lesson from this experiment is this: The analytical proof is achieved by synthesis of a (colored) substance (which, however, is not isolated). This experiment also conveys another important insight into the character of chemistry to first-year students: Because of the necessary preparation, chemistry is very demanding in terms of technical skills. It is, in other words, a systematic art; and the mastery of these technical skills precedes the infusion of theory.

5.3.2 Electrolysis

We know about the composition of water primarily from electrolysis: Placing two electrodes in a sample of water and applying sufficient voltage will produce hydrogen at one electrode and oxygen at the other.[14] This is a synthetic procedure with respect to the products because substances are produced. Only the identification and quantitative comparison of the volumes created make it analytically useful, whereby here I am not including the necessary calibration of quantification. The result is the familiar empirical formula H_2O. The required breakdown of the initial substance water is not yet a sufficient analytic step in and of itself because it does not yield any quantitative information.

5.3.3 Gravimetry

Gravimetry is another prime example of the use of synthetic operations in analysis. In this method, a substance that is not soluble in a solvent is created and weighed, for example barium sulfate in sulfate determination or elementary cop-

14 With a pronounced philosophical accent, Chang (2012) describes how this process proceeds in a manner that is anything but simple and that its history provides fascinating insights into the development of chemistry.

per on a platinum electrode in electrogravimetry.[15] The weighing provides information about the quantity of the target substance. With these methods as well, at least one substance is broken down, that is to say analyzed in the stricter, formal sense of the word. However, this breakdown is by itself not sufficient to provide quantitative analytic information. Chemical analysis is not simply breaking down a specimen into elementary building blocks and then counting these building blocks. Disregarding the information provided by a substance's separability itself, simply breaking it down into individual pieces does not yet provide any reliable information about the original object, even if the building blocks are well known. Like with LEGO®, a pile of pieces yields little to no information about exactly which object was originally present (cf. Figure 3).

5.3.4 Hardness of Water

Even today, titrimetric analyses are crucial to routine laboratory work. Titrimetry evolved during the course of industrial development in the 18th century with the quality control that certain products were thought to require, for example in the textile industry (Szabadvary 1966, p. 213). Titration is a chemical reaction performed under controlled circumstances, usually in an aqueous solution. The target component is referred to as the titrand; the known component added in a defined quantity or concentration is the titrator. In the complexometric titration of "water hardness," the dissolved calcium ions are the titrand and a solution of EDTA (ethylene diamine tetraacetate) acts as titrator. The EDTA "complexes" the calcium ions, and this occurs in the stoichiometric ratio of 1:1, meaning that one calcium ion reacts with one EDTA ion to form a complex. This complex has a certain color, which can act as an indicator of when the endpoint of the reaction has been reached (actually slightly exceeded). In this connection it is important that a synthesis has taken place here too, even if the product is not isolated as a substance. The presence of the product is demonstrated indirectly, and the knowledge of the stoichiometric ratio must be obtained through prior experimentation. The conventional neutralization titrations based on acid-base reactions follow the same principle: In the process known as acidimetry the bases are the titrator, in alkalimetry the acids. The known indicators (such as litmus, methyl orange, and phenolphthalein) help the laboratory staff detect the endpoint of the reaction, i. e., the completeness of the conversion. The crux of the matter with tritrimetric methods in analytical

15 Electrogravimetry is an example of what Rancke Madsen refers to as isolametry. This type of analysis is occasionally referred to as "direct" or "absolute" determination.

chemistry is that they too are synthetic. It makes no difference whether the resulting products can be visualized.

5.3.5 Elemental Analysis

Antoine Lavoisier was the first to employ the combustion of organic substances in a stream of oxygen for research purposes.[16] From one perspective substances are destroyed during combustion while from another products are created. This latter perspective can be utilized in quantitative analytical operations, such as measuring volume or weighing. Justus Liebig (1803–1873) later took up the basic idea of capturing the resulting carbon dioxide in an alkaline solution to quantify the resulting sodium or calcium carbonate by weighing it (Liebig 1837). Liebig constructed the glass apparatus that became famous as the *Kaliapparat* (five-bulb apparatus) and was intensively used in research until the end of the long 19th century.[17] Elemental analysis has remained a routine procedure to this day. The sample is burned and the combustion products (CO_2, NO/NO_2, and SO_2) are determined in the form of suitable chemical products. Here, too, compounds are synthesized for analytical purposes. In Rancke Madsen's system, this would be a *normometric* process.

5.3.6 Helium

The case studies I have discussed so far in my investigation of the relationship between analysis and synthesis in chemistry have been obvious examples. They have dealt with substances that occur in routine chemical practice or experimental laboratory work or at least have a plausible connection to these fields of human endeavor. Let us examine a somewhat more difficult case, that of helium. Inert gases form the final group in the periodic table and under normal conditions do not participate in chemical processes, even in the laboratory.[18] Helium does not readily reveal its chemical propensities because it has practically none to reveal and only occurs in trace amounts on Earth. For this reason the history of its discovery is also rather involved. Just as Bruno Latour describes Pasteur's discovery of the

16 Cf. Szabadvary (1966, pp. 287–309). See also table IX in Lavoisier (1789).

17 Cf. Jackson (2015 and 2023).

18 In the present context it is unimportant that since the 1960s chemists have succeeded in creating compounds of higher inert gases.

lactic acid ferment,[19] the existence of this element was initially controversial. During a solar eclipse in 1868, the French astronomer Pierre Jules César Janssen (1824–1907) observed in a protuberance a conspicuous bright yellow line (wavelength 587.5 nm), which he ascribed to a new element. A few months later, the British scientists Norman Lockyer (1824–1907) and Edward Frankland (1825–1899) confirmed the finding (the latter was initially skeptical about whether helium was an element) and suggested the name "helium" for the new element. The discoverers of helium had no idea that the new element occurred as a gas which is why the ending of the element's name, analogously to discoveries up to that time, suggests a metal.[20] The Italian vulcanologist Luigi Palmieri (1807–1896) detected helium in the lava of Mt. Vesuvius in 1882, and the Scottish chemist William Ramsay (1852–1916) presented experimental evidence of the gaseous state of the target substance and thus of its existence as a substance in 1895. He broke down cleveite, a mineral containing uranium, by boiling it for a prolonged period in sulfuric acid and then studied the mixture of gases produced. Helium was among the substances he detected spectroscopically (Ramsay 1895). However, Ramsay states in his careful concluding remarks that his studies could yet not definitively prove that helium was an element. Nonetheless, the relationship with argon (which is present in significantly higher quantities in atmospheric air and had therefore been found earlier, that is in 1894) and Ramsay's own precision density measurements led him to argue that helium was a chemical element in the modern sense of the word. Ramsay received the Nobel Prize in chemistry in 1904 for his discoveries in the field of inert gases and their classification in the periodic table.

This example provides important insights for the study of the relationship between analysis and synthesis in chemistry. Detecting a single characteristic assumed to be specific, here the line spectrum or parts of it, is not enough to identify a substance. That applies all the more when the knowledge of this substance is fragmentary. Until 1895 not even the existence of helium had been proven, let alone any further characteristics of this substance. The latter could only be demonstrated after suitable preparation techniques such as Ramsay's had been introduced. Only with the presence of comparative samples was further analytical work on this chemical entity made possible and additional information such as density and atomic mass obtainable. The skepticism about whether this inert gas was an element accordingly disappeared within a short time. The source of helium on Earth is, as we now know, the radioactive metallic elements emitting alpha parti-

19 I refer here to the chapter "From Fabrication to Reality" (Latour 2000, pp. 137–174).
20 Bill Brock describes the historical background and provides references for further reading (Brock 2004).

cles. The preparation from minerals described here is admittedly not a manufacturing process in the strict sense, for a chemical element cannot be "made." Nonetheless, the preparation can be seen as a manufacturing step if one understands such a step as manufacturing from suitable initial materials much in the way cooks produce dishes. Regardless of how we choose to describe the process, the same applies to it: Analyzing includes synthesizing operations.[21] In any case, it is difficult to imagine general acceptance of a chemical element without being able to refer to a genuine comparison with corresponding reference material.[22]

5.3.7 Gas Chromatography/Mass Spectrometry

In order to expand support for the main assertion of this study beyond the realm of wet chemistry I will now examine two additional examples in greater detail. Both examples come from instrumental analytics and in modern laboratories are often used in combination (as GC/MS). This practice couples a powerful separation method (gas chromatography, GC) with a highly specific spectrometric detection technique (mass spectrometry, MS). This creates the impression that substances (except of course as specimens) are no longer involved and consequently neither are their manufacture or preparation.[23] This may be why the claim that there is no analysis without synthesis seems particularly surprising in this context. The first example is gas chromatography. In this modern separation method, the sample is converted to the gaseous phase and then passed through a column coated with a separating material, usually a vaporization-resistant polymer. The gas chromatograph has three basic assemblies: In addition to the sample injector, which is attached to the entrance of the column and usually acts as a vaporizer, a detector hanging on the exit of the column registers the passage of ingredients of the sample. The column hangs in an oven with programmed temperature control. The ingredients change between the mobile gas phase and the stationary nonvolatile phase to varying degrees depending on their nature. Under the same conditions, a substance in the same system will always exhibit the same behavior, that is the same retention time (remaining time). A successful separation of substances produces a concentration profile in which the signals ("peaks") are more or less clearly separated from each other. Figure 11 shows such a gas chromatogram in

21 Basically all "pure" substances, even the elements, that we use in laboratories are preparations, see the chapter on Wald (and Figure 4).

22 The methodical and conceptual fringe area of the so-called "superheavy elements" (SHE) cannot be discussed at this point.

23 Cf. the interesting discussion of analytical "dematerialization" in Pierre Laszlo (1998).

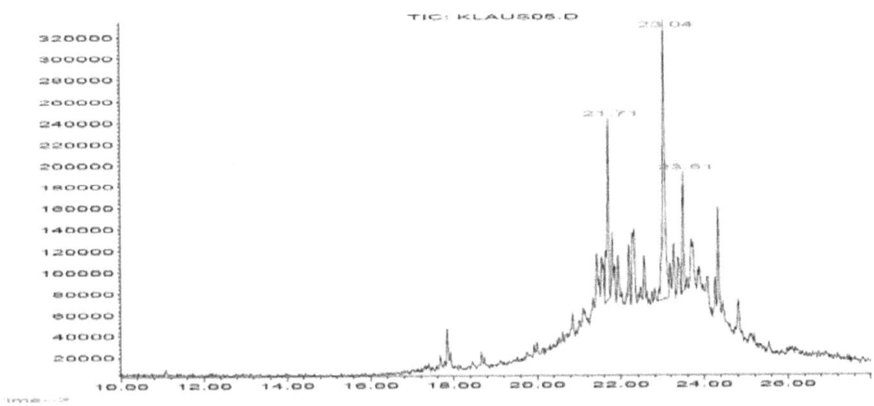

Fig. 11: Total Ion Current gas chromatogram of an archaeological birch bark tar.

which the intensity of the detector signal is graphed over time. The measurement in this case lasted just under 30 minutes.

The sample of this measurement was an archaeological find of birch bark pitch. These materials, which were used for many purposes (adhesive, sealing agent, medicine, chewing gum), have been produced by humans for tens of thousands of years by heating birch bark in the absence of air. The peaks (some are labeled with their retention times) indicate when the respective individual ingredients left the apparatus and were detected.

Within the scope of the present study, one may well wonder what represents the synthetic aspect of this analytical method. I will disregard the fact that the entire separation system including the polymer separating material is itself a manufactured item, i. e., something synthetic in the broader sense. Something else is more important here: Substance transformations occur with the repeated vaporization and retention in the separating material even if conventional jargon terms them “physical.” Substance characteristics, that is “inclinations” of the species under study, are at play here. The ingredients of the sample initially remain chemically unaltered during the separation, yet they change their state. Additionally, the concentration profile that we view as a gas chromatogram, seen in the light of day, has been manufactured. It does not refer back to the intact ingredients. The detection of the separated substances here is not nondestructive.[24] In the case shown

24 Thermal conductivity detectors or even human noses that function “nondestructively” and are actually used for certain lines of inquiry do not come into question in the case under discussion for various reasons. Rather it is the case that the “simpler” detectors most frequently used, the flame

here, the detection was performed by a mass selective detector (MSD), in other words, a mass spectrometer. In such a detector, the molecules leaving the separating system are ionized, and some decay into charged particles. The share of these fragments with positive charges is separated in electrical and magnetic fields (usually a quadrupole) and registered with respect to their time of flight in the MSD. As the time of flight correlates directly with the mass of the fragment, a mass spectrum, that is an ion fragmentation pattern, can be detected. Under defined conditions this pattern is characteristic of the respective type of molecule fed into the system (see below). The Y-axis in the chromatogram shown displays the total ion current (TIC), that is the number of positively charged ions per scan, which in some cases can be several hundred thousand per second. All positive ions detected in one measuring operation are counted together in the TIC. In this mode, the operation of the detector is still nonspecific.

That means that the identity of the more or less separated constituents of the birch bark pitch cannot yet be determined with this information alone. This is the normal case in chromatography. In order to verify which substances are present, one is forced to identify "suspects," provide samples of those, and examine them under the same measuring conditions. If these "suspects" then show the same retention time, then one is at least one step further. Providing the samples includes at least preparation steps, occasionally even genuine syntheses, especially when the target substances are unknown and have to be "cooked up again" before verification is possible. The necessity of providing reference material for calibration in quantitative tasks further expands the need for synthesis in this type of analytics.

With modern data processing technology it is possible to store the mass spectra during a TIC measurement operation and then shortly after detection begin the attempt to identify the constituents of the sample. In this manner one obtains documentation of two characteristics, the retention behavior and the mass spectrum, and the argument based on this circumstantial evidence becomes analytical proof. Figure 12 shows such a mass spectrum. This is the spectrum of betulin, the "biomarker" typical of birch bark derivatives. Betulin is the main component of the white pigment in the outer bark of birch trees. The sample is a different one from the previous one but also comes from the group of birch bark tars. The imaging conditions are the same as with the total ion current chromatogram mentioned above (which contains only very small, scarcely detectable amounts of betulin at about 27.3 min).

---

ionization detector (FID) and the electron-capture detector (ECD), also destroy the sample molecules, that is they transform them chemically, and the measurement is made electrically.

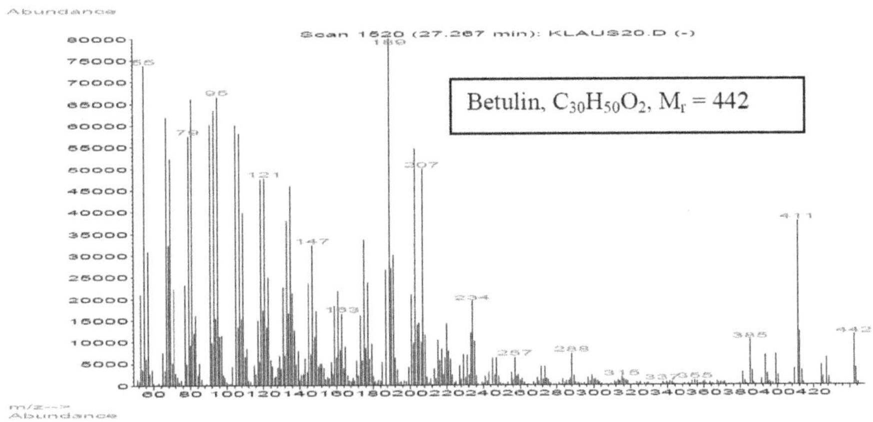

Fig. 12: Electron impact mass spectrum of the “biomarker” betulin from a birch bark tar.

The spectrum shows the ion intensity or quantity per scan on the Y-axis. The X-axis (m/z) shows the fragment mass/charge. Betulin is a naturally occurring substance from the group of the triterpenes. Fragment 442 for example is the molecular cation. All the smaller fragments must necessarily descend from it. This could create the impression that this represents a classic analytical step. Nonetheless, I would argue in favor of categorizing this fragmentation as synthesis in the broader sense, albeit not as constructive joining together. The production of certain chemical entities does not necessarily occur by joining components together. Today, specific programs in conjunction with spectral libraries aid in the identification of substances so that in daily laboratory practice it is not always necessary to synthesize comparative samples. Yet these spectral libraries are based on synthetic work, and they do not contain all desired information. Therefore, it may be necessary to manufacture reference samples for comparison.

In the study of analysis and synthesis mentioned earlier, Claus Jacob refers to precisely the case of the formation of charged molecular fragments we have just examined:

> The expression “analytical$_2$ compound” is not commonly used in experimental chemistry; here it describes compounds that have been generated for analytical purposes (e. g. fragments in mass spectrometry). (Jacob 2001, p. 49)

In using the notation “analytical$_2$” he means “analytical in the experimental sense” in contrast to “analytical in the linguistic sense,” which he expresses as “analytical$_1$.” As was mentioned above, Jacob also mentions synthetic operations for analytical purposes. In the same context in another place he also mentions elemental analysis, already mentioned here, which is a prime example of the synthetic foundation of chemical analysis. This has not been so perceived as Jacob's quotation shows. The expression "experimental analytical compound" sounds nothing short of contradictory to experts because they have become accustomed to overemphasizing abstract submicroscopic descriptions. Yet even in the discourse of the philosophy of chemistry the fundamental synthetic influence on analysis has largely gone unrecognized.[25]

A further aspect of mass spectrometry is easily overlooked. Where does the scale of the mass/charge axis come from? By no means does this scale result from a mathematically formulated physical law; instead it is calibrated with the aid of specific substance specimens.[26]

Although the disciplinary narratives conjure up the "chemical reality of the microcosm" and its form-centric analytical ideal, i.e., the type, number, and arrangement of the elementary particles, there can be no doubt that even mass spectrometry in modern instrumental analytics unavoidably involves synthetic aspects and thus indirectly material aspects as well.

5.3.8 The pH Electrode

A further example from instrumental analytics and the last in this section is unusual in at least two ways. First, it is an instrument that has largely been neglected in the history of science and technology.[27] It is not a large appliance and one whose

25 In her dissertation *Wissenschaft der Synthese – Ein Ansatz zur wissenschaftsphilosophischen Bestimmung der Chemie*, Mareike Frevert studies the five-bulb apparatus and describes it as a purely analytical instrument, thus drawing the relationship between synthesis and analysis toward the conventional side: "Based on this process, it was possible to erect the complex structure of synthesis chemistry" (Frevert 2014, p. 117). From an epistemological standpoint it must not be neglected that in reality, that is in chemical practice, substances such as CO_2 (and from it then K_2CO_3) are created for analytical purposes. Therefore it is useful to take seriously the conceptional separation Jacob suggested and distinguish between linguistic (conceptual) and experimental (empirical) forms of analysis and synthesis. In this manner it appears possible to avoid uncritically confusing the theoretical discussion of analysis and synthesis with the operational steps undertaken.
26 Incidentally the fine adjustment of such a device during operation, "tuning," is performed using a known substance, often perfluorotributylamine (PFTBA).
27 Available information is difficult to locate, scanty, and historically not always neutral and reliable, for example in Stock & Orna (1986) as well as Laitinen & Ewing (1977). The otherwise exemplary standard text by Morris (2002) does not include a single word about pH measurement by means of glass electrodes. Szabadvary (1966, pp. 371–372) provides a brief but very informative insight. The secondary specialized literature is in large part quite helpful, from a historical perspective as well, for example, Dole (1941), Eisenman (1967), Kratz (1950), and Schwabe (1953).

rather unimpressive development occurred independently of the so-called "instrumental revolution" (see Morris 2002). This, coupled with the fact that it is used daily in practically all of the world's working chemical laboratories, has led many experts and cultural researchers to regard it as a boring tool to be taken for granted. Second, the synthetic aspect here is more deeply hidden than in the previous examples.

By far the most common individual measurement in chemical analytics is the measurement of the pH value.[28] The molar concentration (later: activity) of hydrogen ions, expressed since the introduction of the scale in 1909 as a decimal logarithm (pH = – log c [H^+]), is specified as the measure of the acid content of an aqueous solution. In recent years, pH measurement has routinely been performed with the glass electrode. Even the smallest laboratory has at least one such device in its inventory. Such a *pH meter* is repeatedly used in training chemical laboratory and scientific staff. Students of natural sciences learn how to use one in the first year of their studies. These few indications should be sufficient to dispel any doubts about the significance of this device.

In 1906 the German physician and biochemist Max Cremer (1865–1935) published his writings about "electromotoric characteristics of tissue" (Cremer 1906). He begins his extensive essay by explaining his motivation. In so doing he emphasizes that the concentration and nature of ions are sufficient to explain neurologic phenomena without the need for any additional speculative hypotheses. The notion, still popular in certain circles in his time, of a particular vital force inaccessible to instrumental measurement (*vis vitalis*) was alien to Cremer. On the contrary, he became a pioneer of so-called "electrobiology" and therefore of the discipline known today as neurophysiology. He names physical and chemical concepts such as migration velocities and partition coefficients as criteria for the "nature of ions" mentioned, and he proceeds from the applicability of the Nernst equation to biological systems. Although commonplace today, this physical and chemical approach was completely new in biochemistry at the time and was not at all uncontroversial. The Nernst equation brings the electrochemical potential (the "electromotoric force") of a half cell into a law-like mathematically formulated relationship with the type and concentration of the substances used.

Cremer essentially dealt with the electrochemical differences in polyphasic systems ("polyphasic electrolyte chains") as a model of electrical phenomena in living material. In eletrochemistry, one of the main operational problems is the lab-

28 This was the consensus as early as the mid-1960s (Szabadvary 1966, p. 371), yet it has become much more apparent since environmental problems have begun receiving greater attention (see *acid rain*).

oratory separation of "half cells," that is the regions that acting together build up a potential and can thus drive certain biochemical processes. As early as 1890, Wilhelm Ostwald had postulated a theoretical concept for semipermeable membranes for these substance separation operations:

> As is obvious from the previous section, the statement that semipermeable membranes can allow the passage of certain ions and not of others sheds great light on so many and varied fields that one individual can hardly even think of tackling and halfway resolving all of the experimental problems presenting themselves here. (Ostwald 1890, p. 81)

He continues in the same place: "It can hardly appear doubtful that these processes play an important role in the metabolism of the cells."[29]

This is obviously the passage that had a particularly inspiring effect on Cremer, who in his essay takes up one of these "varied fields," which describes the so-called "Pfeffer clay cell" (an impregnated clay beaker with which cane sugar can be separated from water). He then continues:

> Only later was this concept of semipermeability expanded to include those membranes in which individual ions were supposedly unable to pass through the membrane. We owe this expansion to Oswald, and it is also he [...] who should be regarded as the founder of the membrane theory for biological currents. (Cremer 1906, p. 573)

In his 1906 study, Cremer systematically examines so-called "diphasic liquid chains," meaning two-part electrochemical systems. He begins by discussing similar and dissimilar electrolytes in equal and unequal concentrations. In the "primarily experimental section," he reports the results of his experiments, whereby the experiments with glass membranes account for a good fifth of the total length. Cremer had glass spheres 6–10 cm in diameter manufactured with wall thicknesses of up to 2/100 mm. The basic structure of the "glass electrode" can be seen in Figure 13. The two tubes from the inner and outer solutions led to electrodes not shown here. The actual measurement was performed with galvanometers (according to Deprez-d'Arsonval or Einthoven) as was customary at the time.

Soon he experimented with acidic solutions and measured voltages of several hundred millivolts. He formulates his results rather cautiously as follows:

> Thus it seems as if minimal acidity on the one side suffices to awaken great electromotor forces relative to the electromotor forces occurring in physiology. Not only does sulfuric

29 Incidentally, Ostwald himself states that he developed his explicitly anti-atomist stance only well after 1890. At the time of his work on membrane theory, he had not yet rejected conventional atomistic (corpuscular) theory.

acid act this way, but acetic acid [...]. As I introduced into a flask on the outside sulfuric acid diluted to 0.6 percent saline solution and caustic soda solution on the inside, the electromotor force of the combination increased to 0.55 volts, when I otherwise worked with physiologic saline solution. (Cremer 1906, p. 601)

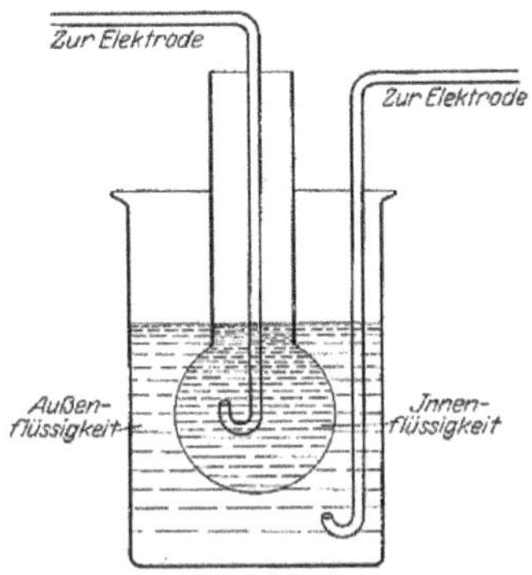

Fig. 13: Main part of Max Cremer's apparatus (from Cremer 1928, p. 1014).

This is the description of the first measurement of acidity by means of a glass electrode. Cremer does not dispute that the electrochemical potential according to Nernst – represented in simplified form as the concentration equalization potential or osmotic pressure – represents the driving force for the buildup of the voltage measured. Nonetheless, the postulated passage of ions through a glass membrane thought to be semipermeable is the main model of his interpretation.

The theoretical penetration of the function of the glass electrode was delayed by two world wars and took many decades as a result. With respect to the actual electrode, it is completely independent of the instrumental analytics that has flourished since the Second World War. The rapid development of semiconductor electronics was of course very helpful for the dissemination of this method, although that had practically no effect on the understanding of how it worked. Since the 1970s a notion has become established which I would like to briefly elucidate below (see Baucke 1994). I assume that the structure of this first and still most important chemical sensor has not changed significantly since Cremer's experiments. Thin glass membranes are still used, although the electrodes have been integrated into so-called "single-rod measurement chains." At first glance, the modern pH electrode looks like a glass tube with a cable connection (which is actually double). The design details are unimportant here, as it is a matter of the occurrence of the electrochemical potential between the inner and outer solutions, that is between the reference electrode and measuring electrode. Glass is a noncrystalline amorphous substance manufactured from three components, a silicate network former (raw material: sand), a network modifier (alkali ions such as Na^+ or K^+ from soda or potash), and a network stabilizer (alkaline earth ions such as Ca^{2+} from chalk). The alkaline ions have a single positive charge, the calcium ions a double positive charge, and the "oxygen ends" of the silicate tetrahedron carry a negative charge. "Modification" means the partial breakdown of the silicate structure (ideally quartz, SiO_2) and consequently the lowering of the melting point to improve workability. "Stabilization" is understood as prevention of the complete transformation into water-soluble sodium silicate (soluble glass). Silicon atoms and oxygen ions are stationary, and the stabilizer ions with their high charge are bound in space more strongly by the oxygen ions. However, the modifier ions exhibit significant mobility within the structure.

Glass has a particular affinity to water. Already Michael Faraday noticed that the electrical conductivity in the surface of glass increases with the degree of moisturization. All glass surfaces that are not in a vacuum form a thin gel layer. Even Fritz Haber, who together with Zygmunt Klemensiewicz was instrumental in popularizing the use of the glass electrode in measuring acidity but categorically rejected the idea of semipermeability, reports this effect. However, he emphasizes that this effect is so small that it influences the application for chemical purposes only in exceptional cases (Haber & Klemensiewicz 1909, p. 409).

Yet it is precisely this gel layer that hydrogen ions can penetrate and which permits the mobile ions in the glass to react to such external chemical influences by changing their position.

Because the modifiers (in modern electrodes usually Na^+ and Li^+) migrate toward the lowest concentration of hydrogen ions, a charge shift occurs and thus a measurable potential between the analyte and the inner standard. The migrations occur very quickly but are not complete. This means that the thermodynamically determined equilibrium is not reached. This measuring process therefore represents a metastable chemical reaction. The silicate structure and practically all of the stabilizers remain stationary, and the movements of the smaller ions are reversible. This means that a properly handled electrode will be ready for use again within a short time.

Although it may sound strange and go completely unnoticed in laboratory practice, the measuring operation with the pH electrode is a synthetic step without a stable product.[30] The metastable interim state allows reliable and repeated measurements, and the electrode can be calibrated for nearly all interesting cases. (The calibration is independent of the phenomenon described and, as has been men-

30 A more elaborated study of the history of the glass electrode provide Ruthenberg & Chang (2020).

tioned, is an additional argument confirming the proposition presented here.) So even in this case, an actual chemical entity is formed to facilitate measurement tasks.

5.4 Further Discussion

This section discusses the central philosophical aspects of this line of inquiry insofar as they have not yet been examined. I would like to note two points prior to beginning this discussion.

First, the *analytic-synthetic ideal* of chemistry has been in effect from the 18th century at the latest to this very day. In my opinion, one could go even further and argue that in experimental practice the precursors of this ideal, admittedly not as formulated concepts but in empirical form, have existed ever since humans have characterized material entities. Here it is unimportant which higher-order objectives the respective researchers pursue and whether they use for the respective operations those concepts that we are familiar with (or choose to recognize). Material individuals were given names based on reproducible empirical experiments long before a general scientific system of substances was developed. Many of these names have performed well in the study of nature for many centuries and some are still used today.

Second (as a reminder): I understand "synthesis" as the manufacturing and preparation character of chemistry in its entire breadth. Chemical processes in which, intentionally or unintentionally, entities arise which were not previously present fall under this concept of synthesis. The conventional interpretation of synthesis as the intentional, planned formation of desired products is included here and has its significance (for example in *total synthesis*) but is not the focus of my study. The preparation of samples for the calibrating comparison in the course of a quantitative analysis is, however, in this interpretation a genuinely synthetic step regardless of whether it can be classified as *isolametry*, *normometry*, or *titrimetry*.

5.4.1 Dialectical Materialism

In so-called *Dialectical Materialism*, analysis and synthesis have an exceptional position. Although the focus here is first and foremost on the theoretical and conceptual meanings of these words, a brief look at this philosophical school of thought is warranted because there was a recognizable tradition of the philosophy of chemistry in the Soviet Union and East Germany (see Laitko 1996) and it can therefore be assumed that the relationship in question had been discussed there. In the Marxist *Philosophisches Wörterbuch* the entry under *Analysis* includes this statement:[31]

> Dialectical materialism regards the goal of analysis not as the simple dissection of a whole into isolated parts. This would correspond to the metaphysical view of analysis. (Klaus & Buhr 1974, p. 69)

The delimitation with respect to "metaphysical" interpretations of analysis is directed primarily against Kant but also against the logical empiricists, who proceed from the strict separability of the two concepts and place greater emphasis on the significance of analysis. Those attacked are indirectly imputed to espouse solely the decompositional variant of analysis. In dialectical materialist doctrine, analysis and synthesis form a whole. The objective of analysis is seen as locating "opposite" components and exposing their "interrelations." The conceptual synthesis, viewed as a higher level of knowledge, emerges as the "union and dialectical negation of thesis and antithesis" (Klaus & Buhr 1974, p. 68).

For philosophical discussions of things like *analytical and synthetic* judgments or *analyticity*, it is fundamentally difficult to find the link to the experimental science of chemistry. Since Aristotle, the concept of analysis in philosophy has focused on conceptual and logical fields. The analogy or projection of *interpretative* or *decompositional* interpretations of "analysis" is plausible but does not say much about chemistry. Nor does the observation that these concepts first appeared in philosophy and that somehow and somewhere they have been applied to or by the science of material substances. The "dialectical" unification of analysis and synthesis mentioned here initially sounds enticing, especially when applied to chemistry. However, it remains vague because what the two terms in themselves mean is never really clarified; and so it is with their relationship as well. The central problem, a fundamental one in the philosophy of science, is to reliably and plausibly link formal theoretical deliberations with practice. Turning the fundamental metaphysical presumption of the dialectical union of synthesis and analysis into a metachemical one requires more than the simple claim that this union is present.

The hope of learning more in general monographs about the postulated dialectics of analysis and synthesis in light of genuine chemical processes is thus largely a false one. However, the more interesting authors in the field of dialectical mate-

31 Comparing the entries in philosophical handbooks and dictionaries, one will notice that as a rule the entries under the heading "synthesis" are significantly shorter than those under "analysis." This corresponds to the relationship of the established disciplines or fields of study "analytical philosophy" and "analytical chemistry" to the lesser established fields on the synthetic side.

rialist philosophy of chemistry provide important observations and food for thought. For example, there is the concept of the "chemical form of motion," which was first postulated by Friedrich Engels in order to systematically organize the sciences. A brief characterization describes this chemical form of motion as a "[...] contradictory union of preservation of chemical elements and formation, transformation, and destruction of chemical compounds" (Laitko & Sprung 1970, p. 44). Assuming that we know what elements and compounds are and how one can acquire knowledge of these chemical species, then the meaning of analytical and synthetic aspects becomes clear. The authors also emphasize that the chemical form of motion can refer to the macroscopic and microscopic realms (Laitko & Sprung 1970, p. 45). This sounds astoundingly non-reductionist for authors with a materialist tenor.

In his study of acids and bases, Rüdiger Simon introduces the expression *synchronous structure*[32] and describes the chemical form of motion in concrete terms:

> Synchronous structures can sway within a range without compromising the identity of a compound [Verbindungseinheit]. Only changes exceeding this range lead to a qualitative leap, to a fundamental alteration of the system. The destruction of the "old" system that this entails simultaneously includes the synthesis of a new one. This process is referred to in chemistry as the chemical form of motion, chemical process, or chemical reaction. (Simon 1980, p. 38)

One can imagine the chemical form of motion as a concrete reaction. Take for example the formation of ammonia from the elements: $N_2 + 3H_2 = 2NH_3$. Here the dialectics of composing (synthesis) and separating (analysis) becomes apparent: Read from left to right, ammonia is created and the original substances "disappear." It is possible to place the reaction and reverse reaction side by side, whereby ammonia would then disintegrate and the original elements would be created (synthesized). The decompositional aspect is only a part of the entire process. If one wanted to determine whether ammonia was an element, then its breakdown would be initial evidence of a compound. However, only the examination of the breakdown products would reveal whether this were actually the case. Yet we must realize that this is initially a purely conceptual interpretation (analytical$_1$ and synthetic$_1$ according to Jacob). As I have attempted to show in the examples, skillful preparation techniques and the manufacture of comparative material for calibration of the necessary quantitative examinations are required in actual laboratory practice.

---

32 Chemical reactions have a time-dependent course. Simon terms the representation of certain aspects of the current or final state "synchronous" and a representation of certain aspects "including the motion" as distinct from the former "diachronic."

The dialectical materialist authors in the philosophy of chemistry largely stick to a formal theoretical and static interpretation. The proposition argued here of an integral production-related aspect of chemical analytics also comes down to a dialectic of analysis and synthesis. However, it does not limit itself to the conceptual and logical interpretation but makes specific reference to empirical findings.

5.4.2 Pessimistic Perspectives for Substance Chemistry?

Our view of material entities has shifted with developments in modern instrumental physical and chemical analytics. The trend toward microphysical interpretations has increased ever more strongly since the early 19th century and thus the disciplinary narratives and definitions have become increasingly further removed from the concept of substance. Joachim Schummer interprets the perspective of substance chemistry pessimistically in light of flourishing instrumental analytics, which in his opinion has superseded the old analytic-synthetic ideal with the "instrumental revolution":

> Once established as independent means of structure determination, spectroscopic methods were also used to characterise quasi-molecular species for which there exist neither a corresponding chemical substance nor a classical approach of structure elucidation [...]. Chemical classification has become more complex and allows more differentiated concepts than hitherto [...]. Species that are defined by spectroscopic properties alone, i. e. by their physical properties, cease to be relational entities as chemical substances and classical structural formulas and thus lose chemical information about reactivities. (Schummer 2002, p. 207)

One historical aspect of this argument is undoubtedly correct: Over the course of the 20th century, instrumental processes for the scientific description of substances have increasingly joined the process of traditional "wet chemistry" so that total synthesis as an analytical tool has taken a back seat. I also raise no objection to the description of chemists' self-image, which has become a matter of course in the wake of the "triumphant success" of atomism and the nearly fetishistic and uncritical embrace of the "unbiased chemical reality of the microcosm," i. e., laboratory jargon. Schummer has undoubtedly correctly perceived the broadening of the field of chemical endeavor with the addition of non-substance species such as intermediates, transitional states, etc. Nonetheless, the "dematerialization" that is occurring predominantly in the conceptual realm reaches its limits in the empirical realm: Material substrates and their reactions are essential where chemistry is actually practiced, regardless of which models or theories are drawn upon for description. I take issue with Schummer's implicit proposition that instrumental measuring processes have displaced the synthetic fundamentals of analytics. No instrumental process, however far developed, can dispense with reference substances for qualitative or quantitative purposes (see spectral libraries, scaling and adjustment, and calibration). This point remains indisputable, even if the focus has shifted away from total synthesis as an analytical tool.

5.4.3 Protochemistry

"Protochemistry" is a part of the research program of the so-called "Marburg School" around the philosopher Peter Janich (1942–2016). The main authors of this program included not only Janich but also his student Nikos Psarros. The researchers regarded as their central task the rational reconstruction of chemistry as a human cultural achievement in express contrast to the predominant understanding of the science of substances as a naturalistic[33] endeavor.[34] The proponents of this school of thought see the driving forces and purposes of chemical activity in practices of the living environment.[35] A statement by Nikos Psarros regarding the topic of this chapter also mentions purposes in a more specified form. In his book about the culturalist reconstruction of chemical methods he writes:

> The use of reactions for trial purposes shall be termed *analysis*, and the corresponding use for production purposes *synthesis*. Accordingly, *analyzing* a substance means obtaining from it known products by means of reactions with (or without) other known initial substances; in contrast, *synthesizing* a substance [means] producing it from known initial substances. (Psarros 1999, p. 127)

This would mean that the purposes determine whether something is referred to as analysis or synthesis. Psarros thus places the cognitive (theoretical) act in the foreground. Let us examine the precipitation of barium sulfate from an aqueous solution as an example, once for qualitative detection of sulfate ions and once to produce barium sulfate. For both cases we write the same equation, that is (greatly

33 Proponents of the Marburg School define "naturalism" as the thesis that "*all occurrence* including human action under historical conditions is a *natural occurrence.* As a research task, the naturalist program demands that all phenomena (regardless of possible differentiations of nature and culture) ultimately be described and explained by natural science." (Hartmann & Janich 1996, p. 14; original author's emphasis)

34 In one of his programmatic essays, Janich demands: "We have to take into account in the theory of science not only the *primacy of practice over theory* and the *primacy of poiesis over empiricism*, but also the *primacy of the physical motion of the craftsman over the determination of chemical qualities of his products.* (Janich 1992, p. 67; original author's emphasis)

35 Substrates of such practices include active ingredients, materials, and fuels (Psarros 1999, p. 27).

abbreviated): $Ba^{2+} + SO_4^{2-} = BaSO_4$. The quoted passage makes correct statements when one takes into account that one time it is sulfate ions that are investigated and another time it is barium salt. The same substance does not have to be referred to for both purposes. And yet this representation has weaknesses. First, not every analytical step is associated with the formation of known substances. For example, the test of decomposability, of whether a substance is an element, practically requires no knowledge of the breakdown products as long as they can be regarded as different from the sample studies.[36] Second, many methods and processes used for testing purposes (examination purposes) do not require chemical reactions. Third, it is usually not practicable to refer in one case to manufacturing and in another to creating (obtaining) products, particularly when one does not know the purpose of the experiment. For the "manufacturing" discussed here, I find it far easier to stick with the meaning of the chemical term synthesis, preferably in the experimental perspective (in accordance with Jacob: $analysis_2$ and $synthesis_2$).

5.4.4 Conclusions

The relation between practice and interpretation in chemistry is dialectical: Chemistry is neither purely analytical nor purely synthetic. Yet this relationship is not symmetrical: Experimental analyses are invariably linked to synthetic operations. Simply separating and quantifying constituents occurs only when one plays with ideas, never in actual chemical practice. Even the test of decomposability, as is required in the Boyle-Lavoisier definition of an element, cannot actually do without the manufacture of a reference sample; the simple breakdown or destruction of a substance only provides the maximum possible information when the breakdown products can be identified. The synthetic operations crucial to analytics range from the preparation of samples to the manufacture of comparative and calibration samples up to and including the synthesis of substances (occasionally as *resynthesis*) in the strict sense.

Historically, the dialectic analytic-synthetic ideal of chemistry developed in the late age of alchemy and in the early modern period. Closer examination of case studies reveals that even in the modern instrumental analytical phase this ideal essentially continues to exist in empirical practice although the impression is occasionally conveyed that a non-material physical measuring principle takes care of

36 It is precisely this point that brought Lavoisier to propagate chemistry as a science that is fundamentally analytic – in the sense of "breaking down."

the "analysis." The verification or falsification of analytical chemical statements therefore proceeds fundamentally, and not only in the realm of classical wet chemistry, by manufacturing suitable substance entities.[37]

---

37 I am very grateful to Tony Edmonds for the initial stimulation to think about the nature of chemical analysis.

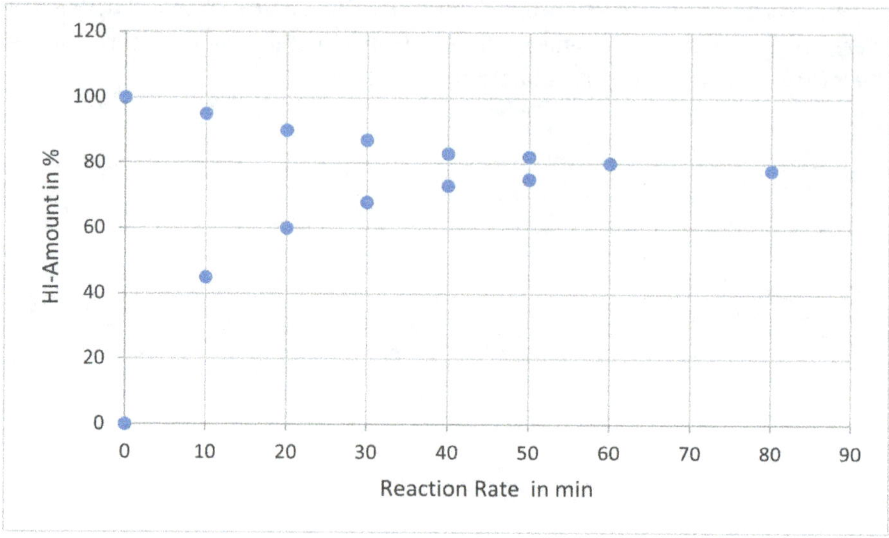

Fig. 14: Course of the forward and reverse reaction (the formation and decomposition) of the process $H_2 + I_2 = 2\ HI$ (at 448 °C, all components in the gaseous phase) reconstructed from original data of the experiment presented by Max Bodenstein (1894) on the formation of hydrogen iodide from the elements. No chemical reaction in equilibrium is complete.

6 Potentiality and Temporality

How do chemical reactions occur? Synthetical chemists are confronted with two problems. The first one is this: Can the desired reaction even exist? Where the answer to that question is yes, the chemist must answer a second question: How quickly and in what way will the reaction proceed? Potentiality and temporality are thus two crucial perspectives in studying chemical processes.[1] Both are involved when the seemingly trivial questions arise as to when the reaction begins and ends. Both are involved when it is deemed necessary to accelerate or terminate reactions. I will devote this chapter to clarifying the question of how potentiality and temporality relate to chemical reactions. I will begin with a discussion of equilibrium and quantity effect in chemistry. I will devote the second section to the historical concept of affinity, the precursor of the current concept of chemical potential, and in the third section I will discuss the epistemological significance of catalysis.

6.1 Equilibrium: Kinetic or Thermodynamic?

In their enthusiasm for his other monumental achievements including quantum theory, people often overlook the fact that in the early phase of his work Max Planck (1858–1947) also devoted himself to the study of relationships that one can unhesitatingly refer to as "chemical." In a series of essays in the *Annalen der Physik* (and *Annalen der Physik und Chemie*) which he began while he was in Kiel, he examined the role of the concept of entropy in substance transformations, a concept which at the time was still new and unusual, especially in chemistry (Planck 1887a–c, 1891). In the third treatise of this series with the subtitle *Gesetze des Eintritts beliebiger thermodynamischer und chemischer Reactionen* he describes equations applicable to reactions in dilute solutions and the gaseous phase. He maintains:

> These essentially confirm the theory of affinity established by *Guldberg* and *Waage*, which thus proves to be a strict corollary of the principle of entropy. (Planck 1887c, p. 462; original author's emphases)

1 I understand "potentiality" as the disposition to realize a certain state and "temporality" as the relation of a process or phenomenon to time.

https://doi.org/10.1515/9783111401041-009

A good twenty years earlier, Cato Guldberg (1836–1902) and his brother-in-law Peter Waage (1833–1900) had published a study with the title *Etudes sur les affinités chimiques* (German version: Guldberg & Waage 1879).[2] This study provided the foundation for the "law of mass action":

> Actually, we understand the active mass to mean merely the quantity of substance within the sphere of action; under otherwise identical conditions the sphere of action can be represented by the volume unit. (Guldberg & Waage 1879, p. 71)

The Norwegian team of authors examined the following general chemical reaction:

$$A + B = A' + B'.$$

This relation can be divided into two parts: the forward reaction (formation reaction) and the reverse reaction (decomposition reaction) if we read in the customary manner from left to right (it also works if read in the opposite direction). The two compete with each other. Empirical findings show that particularly with homogeneous reactions, i. e., when all chemical species are present in the same (better: in the liquid or gaseous) aggregate state (the same "sphere of action"), such a process never leads to the complete disappearance of any of the species involved (cf. Figure 14). In his major textbook *Theoretische Chemie* Walther Nernst (1864–1941) describes the relation as follows:

> Earlier one was often of the opinion that the "reversible reactions" were the exceptions or that one should distinguish two different classes of reactions, the reversible and the irreversible ones; yet such a sharp boundary does not exist at all and there can be no doubt that with the proper experimental setup it can be repeatedly achieved that a reaction will proceed one way one time and the other way the next, i. e., that in principle every reaction is reversible. (Nernst 1921, p. 514)[3]

This simply means that no chemical reaction is complete. Additionally, the significance of the intervention or manufacture by the experimenters becomes clearly apparent here. At this point an ambivalent concept comes into play, the *equilibrium.* Guldberg and Waage explain that equilibrium is present "when the chemical force acting between A and B is equal to the chemical force acting between A' and

2 Regarding the history of the "law of mass action" see Lund (1965) and Nernst (1921, pp. 518–520). The introduction by van't Hoff (1884, pp. 26–36) is also informative.

3 Note that in the German-speaking countries this law has misleadingly been naturalized as "Massenwirkungsgesetz." Because it is about countable quantities rather than about weighable masses, it should better be called "Mengenwirkungsgesetz". Nevertheless, substances remain denoted by non-count nouns (see Chapter 7).

B'" (Guldberg & Waage 1879, p. 71). According to Nernst, Claude Berthollet (1748–1822) first introduced this concept, and he summarizes the Frenchman's basic ideas as follows:[4]

> *The different substances possess different affinity to each other, yet this only comes into play with immediate contact; aside from the chemical affinity, the state of equilibrium also depends crucially on the quantity ratio of the reacting substances.* (Nernst 1921, p. 519; original author's emphasis)

The term "equilibrium" is clearly linked to weighing and scales: Two weights are connected via the balance arm of a scale and gravity causes this arm to move in one direction or the other. In this literal sense equilibrium means the horizontal alignment of the balance arm or the identical height of the scale pans that occurs when two equally heavy weights are placed in the scale pans. In chemistry, two competing chemical states, described by the two sides of a reaction equation, are accordingly placed over the quotients of the mathematical product of the concentrations or, expressed as a formula, $\Pi\ x_i^{vi}$ = const.[5] This constant is called the equilibrium constant (as long as there is no equilibrium one generally refers to the reaction quotient Q). In contrast to the original image of the scale, the concentration ratios in chemical processes do not have to be completely balanced and indeed normally are not. For the conditions present (usually: standard conditions) the equilibrium constant is regarded as a characteristic value for the process in question. Equilibrium therefore correlates with (macroscopic) stasis: One must wait until the scale balances and until the reaction finishes. Yet how do we know whether a substance system is in (genuine) equilibrium or is merely the (false) coexistence of several substances we have placed in a container? There are two conditions: the required components must be present, and they must be able to react with each other. A crucial criterion for verifying the presence of an equilibrium is the sensitivity of the system to "disturbances." Just as a locked scale viewed from a distance appears to be in balance but the second condition is not fulfilled, there are analogous *kinetic impediments* to chemical reactions. A kinetic impediment prevents the reactants from coming together. This is compara-

4 Berthollet, who was active during the transitional period to "modern" chemistry, maintains in his *Essai de Statique Chimique* for example: "Toute substance qui tend à entre en combinaison, agit en raison de son affinité et de sa quantité." (Berthollet 1803, p. 2)

5 The product of the concentrations of the substances x_i to the power of the reaction coefficients ν_i; the latter are in Guldberg and Waage's general example – 1 for the substances on the left side (starting material), for those on the right side (products) +1. Cf. Havemann (1957, p. 125), who formulates this mathematical expression as a condition for equilibrium.

ble to the ancient folk ballad of the two royal children who could not come together because they were separated by a deep body of water. To verify that the system is in equilibrium, one must intervene with a disturbance. With our mechanical example this is relatively simple: Tap one of the two balance arms or one of the scale pans and if the system begins to pendulate, then it was in equilibrium before it was tapped. In a chemical system one can cause a disturbance by adding or removing at least one of the components. If the system reacts with a shift in the concentration ratio (in chemical jargon: achieving equilibrium), then it was in equilibrium before the disturbance. Let us take as an example the formation of ammonia from the elements in the formation reaction $N_2 + 3\ H_2 = 2NH_3$. Under standard conditions, all reactants are in a gaseous state. If after verifying macroscopic stasis one removes some ammonia from the reaction mixture by adding water, which it dissolves into, then the two elements will continue to react to recreate the most favorable product in terms of energy until the original mixing ratio of all "active" components has been restored. Analogously, the opposite would happen if we were to reduce the concentration of one of the elements, it would trigger the decomposition of ammonia. This circumstance is interesting from an epistemological perspective because when verifying an equilibrium by inducing a disturbance, the previous state of the system is invariably detected as the disturbance causes a change in the system. This is due to the relational nature of chemical behavior. The disturbance can be very small and is usually barely noticeable in laboratory work because as a rule repeatability is ensured. Additionally, a chemical system "in motion" can often be examined with minimally invasive methods but that does not change the findings at all: Substance reaction systems must be tested for possible equilibrium by intervening in their substance balance. The concept of equilibrium is thus not self-evident. In any case, "remote diagnoses" are rarely helpful.[6] A system that is macroscopically at rest is not necessarily microscopically or submicroscopically at rest.[7] Forward and reverse reactions cannot be excluded as long as all system components are present:

> If we could monitor an equilibrium mixture at an atomic level, we would find that it is still a turmoil of chemical activity. (Atkins 2015, p. 35)

6 Here it should be noted that the uncritical use of expressions such as "natural equilibrium," "ecological equilibrium," or "global equilibrium" risks turning a statement into nonsense as there may well be no way to verify these concepts.

7 I call those objects "microscopic" which are still discernible under a conventional microscope and those objects "submicroscopic" which are smaller than these and in any case are non-observable.

It was Guldberg and Waage who pursued the plausible notion of explaining the occurrence of equilibrium via the rates of the formation and decomposition reactions, and they were the first to propose the so-called "kinetic" derivation of the equilibrium constant. In the conventional interpretation of this kinetic derivation, which Nernst repeats as late as 1921, the rates of the forward and reverse reactions in equilibrium are equal. Macroscopically, however, the reaction in equilibrium is finished: There are no noticeable or measurable fluctuations in essential characteristics.

Figure 14 shows the courses of the decomposition and formation of hydrogen iodide from the elements in a homogeneous gas reaction at elevated temperatures, which Max Bodenstein (1871–1942) in particular studied intensively.[8] Toward the "end" of the processes the increases in both curves approach zero and finally reach the same equilibrium level. It is clear that completeness cannot be achieved here, neither in decomposition (right → left) nor in formation (left → right). The equilibrium constant has the following form: $K = p(HI)^2\ p(H_2)^{-1}\ p(I_2)^{-1}$.[9] It is crucial to realize how important the design, that is the choice of conditions, is for these experiments. If for example one were to use iodine not in gaseous form but in the solid form that exists at room temperature, then the process would be significantly different although the reaction equation would look very similar: $H_2(g) + I_2(s) = 2\ HI(g)$. The equilibrium constant in this case would have the form: $K = p(HI)^2\ p(H_2)^{-1}$. That means it would depend only on the partial pressure ratio of the two gaseous reactants. It appears at first glance as if iodine were unimportant for the process but merely means that its concentration is not specifiable in the "sphere of action." The iodine would also become a *sink* for freely and dynamically achieving equilibrium, in other words, a kinetic impediment. This generally very interesting aspect is one that Nernst also mentioned, and I will come back to it once again later. I feel another aspect is more important at the moment because it has to do with the reason why I began this section with a reference to a chemistry-related study by Max Planck. The kinetic derivation of the law of quantity action begins, as was mentioned, with formulating the reaction rates for formation and decomposition and equating the competing processes for the case of equilibrium. Proportional influence of the individual concentrations (partial pressures) is assumed for the reaction rates. For the last example mentioned that would be $V_{FWD} = k_1\ p(H_2)\ p(I_2)$ and $V_{REV} = k_2\ p(HI)\ p(HI)$ as well as $V_{FWD} = V_{REV}$ and $K = k_2/k_1$. Both reactions are assumed to be second order, i. e., the sum of the concentrations

8 Bodenstein presented his fundamental results in his Heidelberg dissertation; see Bodenstein (1894).

9 As this is a homogeneous gas reaction, the "active quantities," that is the concentrations of the reactants, can be specified as partial pressures.

of the active substances is 2 for both directions and the result is the equilibrium constant shown above. The problem with this procedure is that far from all "rate laws" can be so easily represented. The derivation applies in the case Bodenstein chose, particularly because of the cleverly chosen reaction conditions, but difficulties arise even with the change in the aggregate state of the iodine mentioned above. The matter becomes really complicated if one examines the superficially analogous reaction of the formation and decomposition of hydrogen bromide, HBr.[10] The kinetic derivation of the equilibrium constants is thus not universally applicable. At best it can serve as a plausibility check.

In his multi-part study of the principle of propagation of entropy, Max Planck does not follow a kinetic path but that of phenomenological thermodynamics. From this perspective, entropy is defined as the ratio of the reversibly conducted quantity of heat to absolute temperature ($S = Q_{rev}/T$). Planck states that a substance system is in equilibrium when maximum entropy is reached (in other words, when $dS = 0$). Taking the thermodynamic route, he arrives at the same situation that Guldberg and Waage describe and that van't Hoff (1884) also discusses in his major study on kinetics. Planck also encounters "apparent deviations from the theory" with respect to the vapor pressure and the lowering of the freezing point in salt solutions and concludes:

> Every apparent deviation from the theory is an indication that the molecules of the salt in the solution are not those that one ordinarily regards as such but that the salt in the solution has undergone a chemical change. (Planck 1887c, p. 501)

With this statement the thoroughly informed theoretical physicist confirms the priority of empirical facts and, not insignificantly for the line of inquiry pursued here, the fundamental significance of thermodynamics for chemistry. The fact of "electrolytic dissociation" is what is behind the "chemical change" he refers to here: A salt in solution dissociates, and when one previously expected a certain substance, say, table salt (NaCl), in solution it becomes two species (Na+ and Cl^-, both hydrated).[11]

In the newer textbooks of physical chemistry, the connection between the chemical equilibrium and the concentration of the substances involved is derived from entropy, more precisely from the chemical potential. The chemical potential

---

10 The rate of formation of hydrogen bromide is $V_{HBr} = k_1\, p(H_2)\, p(Br_2)^{3/2}\, (p(Br_2) + k_2\, p(HBr))^{-1}$. The reaction order, which is often around 1 or 2, cannot be specified with a simple number for this radical chain reaction; cf. Atkins & de Paula (2013, p. 281).

11 In (submicroscopic) reality, ions should be regarded as separate even in solid bodies and in solvation (in water: hydration) are only separated out, not newly formed. Note that Wilhelm Ostwald got into severe trouble to understand ions without accepting particles, see Ruthenberg (2008a).

is identical to Gibbs' enthalpy or "free" enthalpy. The latter can be relatively easily deduced. First, one examines the entire substance change (entropy change) as the sum of the entropy changes in the reaction system and the environment:

$$\Delta S_{tot} = \Delta S_{sys} + \Delta S_{env} \qquad (1)$$

The environmental entropy is influenced by the reaction system, i. e., if the reaction gives off heat to the environment, then the entropy of the environment increases (the exothermic case) or, in the opposite case, decreases (the endothermic case). One can therefore convert equation (1) to the following equation by substituting, according to the definition, the change of the environmental entropy with the quotient of the reaction enthalpy and the temperature (that is by introducing the first law of thermodynamics). To correctly specify the direction of the process according to the customary convention, the sign of this quotient must be reversed:

$$\Delta S_{tot} = \Delta S_{sys} - \Delta H_{sys}/T \text{ with } p,T = \text{const.} \qquad (2)$$

Expanding this with –T yields

$$T\Delta S_{tot} = -T\Delta S_{sys} + \Delta H_{sys} \qquad (3)$$

Both sides of equation (3) now have enthalpy (energy) character. The term on the left side has since the first derivation been customarily called "free enthalpy." In honor of the two researchers who made major achievements in this field, the slightly altered equation is occasionally known as the Gibbs-Helmholtz equation or simply the Gibbs equation:

$$\Delta G = \Delta H - T\Delta S \qquad (4)$$

Free enthalpy (Gibbs enthalpy) is the crux of thermodynamics. In a certain sense it links the first and second laws of thermodynamics.[12] If $\Delta G < 0$, then the process in question is spontaneous ("voluntary") and is termed "exergonic"; if $\Delta G > 0$, then the process cannot take place under the given conditions and is termed "endergonic." At $\Delta G = 0$ the system is in equilibrium.

When appropriate data are available, free enthalpy determines whether a chemical reaction is possible (under given conditions). Consequently, a generally valid kinetic derivation of the law of quantity action is not possible (the time value does not occur in the equations listed); instead, the law must be deduced thermodynamically, i. e., via entropy. Guldberg and Waage were perfectly correct

---

12 Unlike Planck's understanding of it, entropy for the followers of energetics is a construction derived from energy without any illustrative correlate. František Wald attempts to expand this notion in his short work with the expressive title *Die Energie und ihre Entwertung* (Wald 1889). Wilhelm Ostwald in his *Grundriss der allgemeinen Chemie* dispenses with the term "entropy" entirely.

in referring to the "chemical forces" of the competing forward and reverse reactions, but it was only Planck and others who followed the generally applicable path in interpreting chemical equilibrium.

"One of the most important [...] relationships in chemical kinetics is the equation that links the rate constant of a reaction to temperature." (Laidler 1970, p. 49) This equation may be found in current textbooks and handbooks under the entry "Arrhenius' equation." It has the form $\ln k = \ln A - E_a/RT$, with the pre-exponential factor A, gas constant R, absolute temperature T, and activation energy E_a. Svante Arrhenius derived this equation from Jacobus Van't Hoff's (1852–1911) approach for the temperature dependency of reaction enthalpies (Arrhenius 1889). With respect to the philosophy of science it is interesting because the "kinetic approach" was used for it as well, an approach whose underlying assumption (that the reaction orders of the formation and decomposition reactions are equal) does not universally apply, as we have seen. If the results of rate measurements do not follow Arrhenius' equation,[13] then this is a clear indication that the process proceeds in a more complex manner than a first or second-order reaction. There is thus no universally applicable theoretical connection between kinetics and thermodynamics in chemistry.

6.2 Affinity

The theory of affinity has long been a part of chemical research. It is essentially its very heart. In this section I will examine some historical precursors of what now is called chemical potential.

6.2.1 The Strength of the Acids

One typical concept from the field of chemical potential is that of "acid strength." I will examine it in this section. The world of real-life experience played into the choice of words: The strong defeating the weak was certainly a vivid image that came to mind. In a greatly simplified sense acid strength can be correlated to the active concentration of the respective compound. However, this does not get us very far because the same concentration can be achieved with different acids. One can also attempt to rate the strength of acids according to their ability

13 This occurs graphically by plotting ln k over 1/T; if a straight line with a negative slope results, then one can easily determine A and E_a.

to dissolve precious metals. At first glance it would seem plausible that the strongest acid is that which can dissolve the most precious metal. For example only aqua regia, a mixture of hydrochloric and nitric acids, is capable of dissolving gold; and nitric acid dissolves copper, which sulfuric acid and hydrochloric acid cannot. Acid strength can just as easily be represented via another relational matrix, namely the so-called "displacement reaction." In such a process, the strong acid "displaces" the weaker one from its compound (its salt). That is to say it releases the weaker acid, as in this example in modern notation:

$$HCl + CH_3COONa = CH_3COOH + NaCl.$$

This type of description is obviously independent of the current concentration of acid, rather it refers to a substance-specific disposition, namely the relational behavior of the entities in question. This behavior is interpreted as the capacity of more complete dissociation, that is the more effective cleavage of hydrogen ions.

Thinking in relations – naturally initially without any atomist interpretation – increasingly gained acceptance at the latest with the chemical theories of affinity of the 18th century. For example, Etienne-Francois Geoffroy (1672–1731) published his very influential affinity table in 1718. It includes the ranking: sulfuric acid, nitric acid, hydrochloric acid, and acetic acid. That means: If one adds hydrochloric acid to the alkali salt of acetic acid, then one will release the acetic acid and alkali chloride will form (as in the equation above). The first ordinal scale of acid strength was thus created. The level of knowledge regarding acid strength remained at this point for some time. Then Wilhelm Ostwald appeared on the scene and launched his scientific career in an explosive and experimentally very creative manner. Long before an explicit and quantitative acid-base theory became established and long before the introduction of the now familiar pH concept, he set about quantifying the concept of affinity, which until then had been purely qualitative. He devoted the greater part of his doctoral dissertation with the title *Volumchemische und optisch-chemische Studien* to acid strength. Two goals motivate him in these studies. The first of these goals is the "volume-chemical" measurement of relative affinities; the second, and for him more fundamental, goal is the study of the state "of the salts in aqueous solution and the chemical function of the water of the solution" (Ostwald 1878, p. 3). For his series of experiments he uses a glass apparatus he himself developed. It consists of a graduated pipette in which neutralizations are performed and the resulting dilatometric effects can be read off very precisely on the solution column (Ostwald 1878). It was known that the volume of the reaction solution changes in reactions between acids and bases. If potassium hydroxide or sodium hydroxide (potassium hydroxide or sodium hydroxide solution) are used as bases, then an increase in volume will result, and with ammonia a decrease in volume.[14] Ostwald's idea was to quantify these effects. His results are consistent with Geoffroy's qualitative conclusions as well as with the modern data for acid constants with one not unimportant exception. Ostwald norms his measured results and assigns what he feels is the strongest single-base compound, nitric acid, the value of 100. Hydrochloric acid, from a current perspective "stronger" than nitric acid, is assigned the value 98, acetic acid 1.2. The current table values for the acid constants, i. e., the equilibrium constants for the processes of the general form $HL(aq) = H^{+}(aq) + L^{-}(aq)$ are (in Mol/L) nitric acid/$10^{1.37}$, hydrochloric acid/$10^{6.1}$, and acetic acid/$10^{-4.76}$. That means that hydrochloric acid from the current perspective dissociates significantly more strongly than nitric acid.[15] Ostwald does not comment on the acid-base theory he may espouse or assume and logically does not offer any interpretation of the substance behavior he quantifies. And basically he does not require any such theory for his purposes. It was only in 1887 that Svante Arrhenius publicized his theory of electrolytic dissociation on which the modern "proton" acid theory is based (this, however, was only explicitly formulated in detail in the 1920s). In his later *Grundriss der allgemeinen Chemie* Ostwald discusses the dissociation coefficients of acids ($k = a^2/(1\text{-}a)\ v$, with a as the dissociated proportion and v as the volume) and concludes the following:

> The formula contains a single constant k, which is determined by the nature of the substance, and this constant is the sought-after measure of chemical affinity. (Ostwald 1899, p. 524)

Ostwald is well aware of the historical background of chemical affinity. He emphasizes that the beginnings of the formulation of affinity tables had to do with the relatively easy observability of salt reactions and names Stahl, Geoffroy, and Torbern Bergman (1735–1784) as early researchers of chemical affinity. He maintains that these researchers' affinity tables were discarded because they were based on "insufficient premises" and describes Berthollet in particular as the instigator of the disappearance of these tables.

> They [the insufficient premises] were based on the view that affinity was to be understood as a sort of mechanical force. Just as a greater force "overcomes" a lesser one so that the bodies

14 Laidler's (1993, p. 279) conclusion that Ostwald found in his master's thesis and doctoral dissertation that the acid-base affinity does not depend on the nature of the base is not correct.

15 The fact that Ostwald did not notice the difference of about five orders of magnitude in the degree of dissociation of these acids may be attributable to the increased concentration of his experimental setup (probably in the molar range). At high concentrations, the ions interfere more strongly with each other so that the effective concentration of hydrogen ions (or affinity) is significantly less than assumed.

move according to the greater force, chemical reactions were also thought to occur according to the greater affinity. (Ostwald 1899, p. 523)

Ostwald describes the blossoming of organic chemistry in the 19th century with its primarily structural and theoretical bias as the main impediment to affinity research and emphasizes the significance of the work of Guldberg and Waage as a promising revival of this field.[16]

6.2.2 The Systematic Art of Reaction Control

Not only the characterization and classification of the substances but also their manufacture naturally proceeds from the focused practical interaction with their relations and dispositions. Many substances can be formed in different ways as can be illustrated by the so-called "water traps." In these glass apparatuses that are routinely used in esterification reactions (reactions of carboxylic acids with alcohols), water is produced as a "byproduct" (and more importantly: an energy source). Many do not regard the formation of water in the procedures described as synthesis in the conventional sense. However, this is unimportant for the current discussion. Rather what is important is that over the course of history the experts in the science of substances have learned to use their inventory of substances and technical equipment ever more ingeniously to manufacture known and new materials. In the 19th century at the latest, there was a great surge in the focused synthesis of organic compounds in particular. Against the background of the previously discussed dual research approach of physical chemistry, kinetic vs. thermodynamic, one question naturally arises: In what way can we really influence the success of synthesis? A common response among experts is this: There is kinetic reaction control and thermodynamic reaction control. Some authors accordingly refer to *kinetic* products and *thermodynamic* products of chemical reactions. What do they mean by this? Let us look at an example. If a system is in equilibrium, then according to the thermodynamic interpretation the most stable of the alternate products is the one which is farthest into the exergonic range. This is how one explains why when carbon is burned more carbon dioxide (CO_2) is formed than carbon monoxide (ΔG_{CO} = −137 kJ/mol; ΔG_{CO2} = − 394 kJ/mol) if all reactants come together. If the kinetic impediment mentioned comes into play, for

16 It should be noted that even the title of the periodical established by Ostwald and van't Hoff in 1897 speaks for the independence and significance of affinity at this time: *Zeitschrift für Physikalische Chemie, Stöchiometrie und Verwandtschaftslehre* [Journal of Physical Chemistry, Stoichiometry, and Affinity Theory].

example, if there is insufficient air supply, then larger quantities of the thermodynamically disadvantaged carbon monoxide (CO) can form. In lab jargon, one would call CO_2 the *thermodynamic* product and CO the *kinetic* product. However, it should be noted that the driving force behind both partial processes is energetic, that is thermodynamic, in nature. In this regard kinetics and thermodynamics are not in a symmetrical relationship: There is no kinetic causation. The more stable product represents the (apparent) end of the reaction sequence, yet it need not be that which we can observe macroscopically. For example, the snow on the ground from yesterday's snowstorm is not at all stable from a thermodynamic perspective if the temperature has since risen above 0 °C. Under these conditions one can better describe its state as *metastable.*[17] Reaching equilibrium takes time and precisely this is the crucial point of the discussion: The term "kinetic product" is misleading. The substances to which some apply this term owe their existence to a kinetic impediment, yet they are caused by chemical potential, that is to say thermodynamically. This is hardly a trivial fact for the philosophy of chemistry. Without this phenomenon there would be no life: If chemical equilibriums were reached immediately the world would be rather boring from a substance perspective. Yet another aspect is of interest in this chapter: What exactly does reaction control involve? The first way to influence the result of a planned synthesis naturally involves the selection of starting components and the reaction vessel. The selection of the reaction matrix, for example the solvent, is not unimportant (we will disregard possible catalysts for the moment; their importance will be discussed in the next section). Once the starting components have been selected, the thermodynamic endpoint of the process is determined by natural law, when equilibrium is reached. If a substance is *endergonic*, that is it has a positive free standard formation enthalpy, then it can still be created if an *exergonic* synthesis path can be found for it. A familiar example is ozone (O_3), which is thermodynamically metastable ($\Delta G = 163$ kJ/mol), that is under standard conditions it cannot be formed from "even" oxygen (O_2) and yet it is measurable in the Earth's middle atmosphere and, although present only in ppm traces, it has a life-preserving function, namely the absorption of ultraviolet light. If we reverse the perspective it becomes clear that the decomposition of ozone is thermodynamically determined.

The thermodynamic basis of a chemical process cannot be manipulated at will. If one intervenes in a chemical equilibrium by changing the concentration of starting materials or products, the concentration ratio will change automatically ac-

---

17 Metastability, which I discussed in the chapter on Ostwald, is a commonly encountered form of a material-related kinetic impediment: The low carbon content in steel is another familiar example: At equilibrium there is no steel.

cording to natural law (see Bodenstein's experiments and the synthesis of ammonia). The thermodynamics in such cases remains the same.[18] It is possible to actually *shift* the equilibriums of endergonic processes by increasing temperature: When "burning" lime ($CaCO_3$) to form "burnt lime" (CaO), a starting material that is stable under standard conditions is converted into the desired product.[19]

Catalysis, that is steering the reaction paths and thus the reaction rate by influencing the reaction matrix, is a fundamental part of the systematic manufacturing practice of chemistry. It is the topic of the following section.

6.3 Catalysis

Strangely, the recent philosophy of chemistry lacks any detailed studies of catalysis, a process whose technical applications continue to greatly affect our daily lives and which has been researched at enormous cost for more than a century in the chemical sciences. This deficiency cannot be remedied in its entirety here. However, I would like to draw the reader's attention to a few crucial aspects. The main question will be: Actuation or acceleration?

I mentioned slowing down or preventing chemical reactions in my discussion of the *kinetic impediment*. Whereas the art of experimentation does not have unlimited influence on the result of synthesis, one can affect the path and rate of the reaction by manipulating the reaction conditions. Slowing down a reaction (in the sense of *negative catalysis*[20]) is always possible in principle. But can the opposite be achieved as well? How is it possible to accelerate a reaction? The fundamental idea is obvious: If the distance between the reactants slows down the reaction, then they must be brought together more often per unit of time in order to accelerate the reaction. The collision of particles of the reactants is a prerequisite for realizing the reaction. Whereas, as we have seen, chemical thermodynamics large-

---

18 In contrast, Statham (2017) speaks misleadingly of "thermodynamic manipulation" and states: "It is therefore possible to push the reaction in either direction (i. e. towards or away from the thermodynamic product) by adding or removing product or starting material. It is this process that is exploited in Fisher [sic] esterification: by either using an excess of the alcohol (one of the starting materials) or removing water (the byproduct), the reaction is driven towards formation of the ester." (Statham 2017, p. 4826) The author describes here the ester formation and water separation mentioned in this section.

19 When examining Gibbs' equation ($\Delta G = \Delta H - T\ \Delta S$), it becomes clear that the increase in temperature increases the entropy term. Increasing temperature in exergonic processes can influence the equilibrium only unfavorably.

20 Schwab (1931) argues convincingly for including the concept of impediment in the theory of catalysis.

ly does not require submicroscopic images, modern chemical kinetics, that is the field in which temporality is the central notion, immediately launches conceptually into the submicroscopic range. And yet the empirical starting point, as always in chemistry, is the phenomenology of substance. Of the four state variables pressure, volume, temperature, and quantity of substance (which by the way are closely linked to the concept of substance), temperature has the best chance of influencing the reaction. A rule of thumb is that a temperature increase of 10° will double the reaction time. By warming a reaction mixture while taking care not to overheat it (which would shift what under normal conditions would be an exergonic reaction into the less favorable range of the opposite reaction), it is possible to accelerate the reaction to a certain degree.

The category of time for describing chemical reactions was introduced relatively late, in the second half of the 19th century. The more precise study of reaction rates and courses, i. e., the emergence of the branch discipline of chemical kinetics, brought with it the development of notions about the processes we call "catalytic." Catalysis in chemistry describes a phenomenon whereby a reaction is accelerated by a substance that itself does not occur in the reaction equation. Wilhelm Ostwald, honored with the 1909 Nobel Prize for his work in precisely this field, aptly summarizes the phenomenon nearly a decade later:

> In many cases the rate change from the presence of such substances is slight, in others very significant; in certain cases even almost immeasurably small amounts of foreign substance are enough to increase the rate to several multiples of its initial value or reduce it to a small fraction of it. (Ostwald 1917, p. 327)

The term was introduced in 1836 by Jöns Jakob Berzelius who identified the phenomenon (cf. Laidler 1986). He describes it as different from conventional chemical affinity and suspects that its nature is not dissimilar to the known electrochemical relationships. As so often in chemistry, where the success of an application (which by the way correlates with the Berthelot principle) has the highest priority,[21] the term and even initial applications had become accepted long before the infusion of theory. The more familiar early applications of technical significance include the production of sulfuric acid from a sulfurous acid using nitrous acid or nitrogen oxide (the lead chamber method), acid-catalyzed reactions in organic chemistry, the breakdown of starch into sugar in the presence of acids, the oxidation of combustible gases on heated platinum, and also the oxyhydrogen reaction on the so-called "platinum sponge" at room temperature, which became famous with *Döber-*

21 In medicine there is the proverb "Whoever heals is right." In chemistry this would be analogously "Whoever can manufacture the product is right."

einer's lamp. What these reactions have in common is that certain substances influence the process without themselves being consumed. It was soon noticed that this phenomenon is very specific, that not every substance can influence any random reaction. However, there was no "catalytic affinity theory." Berzelius gave the phenomenon a name but made no attempt to postulate a theory. Nonetheless, he spoke of an "intrinsic force" and suspected that giving it a name would be helpful to further research:

> Making use of a well-known derivation in chemistry, the *catalytic force* of the bodies, I will therefore call breakdown by this same thing *catalysis*, in the same manner as we understand, by virtue of common affinity, the word analysis to mean the separation of the components of a body. (Berzelius 1836, p. 243; original author's emphasis)

As early as 1834 Eilhard Mitscherlich (1794–1863), who had been with Berzelius as a post-doctoral student and since 1822 was a professor in Berlin, had introduced the expression "contact reaction" for describing the acceleration of a reaction.[22] Berzelius criticizes the latter with the argument that contact between substances is necessary for all chemical processes and the term "contact reaction" is therefore not sufficiently specific. Yet the term "catalysis" also has its drawbacks. The German translation *Zersetzung* [decomposition] includes the particularly interesting synthetic aspect of such a process as little as it does the acceleration phenomena. Additionally, a literal interpretation places the term overly close to the decompositional meaning of "analysis." Thus it is more the power of personal influence and not the intrinsic conceptual power of persuasion that is behind the general acceptance of this expression. Berzelius found a worthy opponent in the person of Justus Liebig (1803–1873), who refused to acknowledge the systematization under the name "catalysis" and in particular flatly rejected the term "catalytic force" as nothing more than mere speculation. In it he saw the introduction of a natural force in addition to those recognized at the time such as gravity and the electrostatic force that was seen as essential for chemistry, due in no small measure to its propagation by Berzelius. Friedrich Wöhler invested a lot of effort and time in explaining to his friend Liebig the ideas of his Swedish teacher and in attempting to turn him away from the reproach of a new *vis occulta* (after the *vis vitalis* which was thought to be overcome), yet his efforts met with little success.

Ostwald later stood by Berzelius, although he himself missed no opportunity to criticize theoretical speculations poorly supported by empirical findings and in his own work usually tended not to venture into theory, as with the affinity studies early in his career. In *Grundriss der allgemeinen Chemie* he notes:

---

22 Regarding Mitscherlich and his occasionally underrated achievements cf. Schütt (1992).

> The name given [catalysis] had, as always in such cases, the purpose and benefit of defining the task with greater clarity and thus preparing its solution […]. Meanwhile, this gain has been nearly completely destroyed by a hypothesis postulated by Liebig. Liebig assumed that a substance in decomposition or "chemical motion" can communicate its "chemical motion" to another that is present and not decomposing and thereby bring that substance to decomposition. This view is based on a mere analogy; yet it was viewed by contemporaries as more scientific than the classifying act performed by Mitscherlich and Berzelius and because of the ensuing polemics the expression catalysis remains, scientifically speaking, somewhat shabby even to this day. (Ostwald 1899, p. 518)

According to Ostwald, the "apparent explanation of the instigation hypothesis [of Liebig's] […] has had no success other than to delay the study of the problem for half a century" (Ostwald 1899, p. 518). Liebig finds the idea of a "force" inappropriate and replaces it with another speculative notion which presumably comes from comparison with fermentation processes.[23] Ostwald assumes that the cause of a substance change is a chemical potential, that is in thermodynamic terms negative free enthalpy. Consequently, he feels the instigation hypothesis is not conclusive. In his view, a chemical process is already underway, albeit immeasurably slowly.

Alwin Mittasch (1869–1953), who studied under Ostwald and received his doctorate under Bodenstein, argues in his writings on the philosophy of chemistry for another interpretation.[24] For him the catalytic effect is an occasioning, initiating, or actuating causality. In his writings we find a scientific and methodological standpoint that is extremely rare among chemists:

> The classification of catalysis within the study of reaction kinetics is a matter of science; its epistemological classification within the whole of nature's causal relationships is a matter of philosophy. Yet it is noteworthy that this philosophical work has primarily been undertaken by men who were natural scientists and not specialized philosophers. (Mittasch 1948, pp. 268–269)

Mittasch does not give very good marks to the community of specialized philosophers of his day when it comes to "chemical" or "catalytic" causality.[25] However, this is hardly surprising because in his day and even today there are very few philosophers who had or have developed enough interest in chemistry as a reference science. And therefore a few chemists have "of necessity" become active in the phi-

23 Cf. Mittasch & Theis (1932, pp. 94–117) .

24 For a brief outline of Mittasch's philosophy of chemistry see Ruthenberg (1996). More detailed yet ideologically tendentious is the section devoted to Mittasch in Simon et al. (1982, pp. 155–160). See also Ruthenberg (1995b).

25 Mittasch discusses only Hans Driesch (1867–1941), who comes from biology, in a longer passage from his concepts and opinions of nature.

losophy of chemistry themselves. It is also interesting to note that the research interests among the handful of academic philosophers who were initially chemists usually have little to do with chemistry.[26]

The (indirect) dispute between Mittasch, the proponent of the catalytic actuation thesis, and Ostwald, the "thermodynamicist," is not merely one of words. Ostwald rejects the causal influence of the catalyst as a driving force categorically; Mittasch, who played a crucial role in developing the catalytic synthesis of ammonia and therefore very well knows what he is talking about, does not mean only his actuation thesis. Rather he would like to consider both elements, the driving force and the actuating effect, as an integrated whole. As we have already seen, no substance permutation is actually complete. A piece of brass for example is regarded as "solid." Yet we can smell it if we hold it close enough to our nose. That means that a tiny portion of the metal is not in a solid state but in a gaseous state. The notion of an infinitesimal reaction rate prior to contact with the catalyst thus cannot be excluded. Ostwald is not wrong in this regard but he does not see the whole process. Let us consider a non-chemical example: a brick lying on the edge of a roof. The potential that "pulls" the brick down can be easily quantified if we multiply mass, height, and acceleration due to gravity. And by equating this potential energy with kinetic energy, we can also predict the speed at which the brick will hit the ground when it falls. Thus we would "understand" the cause and even some of the future events. Yet the crucial point that Mittasch describes better in the corresponding chemical case is this: We do not know if or when the brick will leave the edge of the roof. In essential ways it is the same situation that we face when bringing about chemical equilibriums. Comprehending potentiality is necessary but not sufficient for a comprehensive description and explanation of the process. The actuating or activating element is an integral part of the description of temporality. Let us return to our example: Whoever accidentally or maliciously kicks the brick and thus sets the process in motion (releases it from its impediment) is part of the interpretation. And to speak here of an actuating causality as Mittasch understands it does not necessarily include the claim of causation. One can conceptually separate the actuation and the reason for a (chemical) process. It is surprising that even after having received Nobel honors, Ostwald is not yet clear about the nature of catalysis here in the fifth edition of his *Grundriss der allgemeinen Chemie:*

> Catalysis has not yet been scientifically explored in great detail because, despite its importance and despite its having been known for about a hundred years, it has only recently

26 Chemists who later became professors of philosophy include Hans Cornelius (1863–1947), Grover Maxwell (1918–1981), and Werner Leinfellner (1921–2010).

> begun to be studied systematically [...]. As regards the previously accepted laws of catalysis, it should first be emphasized that this is a very general phenomenon. There is presumably no reaction whose rate could not be influenced catalytically and no substance that could not act catalytically in some reaction. Yet no general law has yet been found that expresses specific relationships between the type of reaction and that of the catalyst [...]. Thus elements that exhibit multiple readily changeable oxidation states often act as oxidation catalysts, and chlorinations are accelerated by appropriate chlorides or their basic substances [...]. Meanwhile precisely these cases have suggested a theory which in many cases has proven in essence to be correct. It consists in the assumption of an intermediate reaction [...]. (Ostwald 1917, pp. 328–329)

Here we are reading the writings of a Wilhelm Ostwald who by his own standards is leaning pretty far out the window. For a long time he applied his eloquence and influence in fending off speculative theoretical entities. Here we encounter him in what could be called a transitional stage on the way to modern atomist chemistry. The notion of intermediate reactions or even intermediate stages that one can detect empirically only indirectly or not at all is something he would have rejected categorically earlier. However, in today's chemistry in which theoretical and manifest entities are more or less treated equally the so-called "theory of the transitional state" is the accepted model. As was mentioned, Ostwald prefers the acceleration concept to the actuation concept which is why he does not use an expression such as "activation energy."

Actuation cannot occur without expending energy. Kicking the brick may cost only a small amount of energy but it is a necessary amount, although it does not appear in the energy balance of the process in question. This line of thought also makes it clear that the dispute between Ostwald and Mittasch is attributable to the difference in how they view the system. Ostwald assumes a closed system (an inertial system) that allows an exchange of neither substances nor energy; Mittasch's conception allows the controlling intervention of the catalyst by "contact" from outside.

The idea of the contact effect in catalysis seemed plausible for the cases of actuation, for example, the oxyhydrogen gas reaction upon contact with a platinum surface (see Döbereiner's lamp) and similar processes. Accordingly, this had been discussed for some time, very prominently by the Basel chemistry professor Christian Friedrich Schönbein (1799–1868) in his treatise *Über die Häufigkeit der Berührungswirkungen auf dem Gebiete der Chemie* of 1844.[27] Whereas Ostwald in his own studies almost exclusively observed homogeneous systems, that is gases and solu-

27 See Schönbein (1844). Regarding Schönbein's work on catalysis see also the ninth chapter in Mittasch & Theis (1932, pp. 143–166). Benjamin Steininger (2017) provides an interesting link between catalysis and the history of media and material.

tions,[28] Mittasch at BASF had researched and developed the heterogeneous catalysis of gas reactions by solid bodies (especially synthesis of ammonia) for many years. These different lines of work doubtlessly had no small influence on these researchers' choice of their respective theoretical interpretation. The actuation conception correlates very well with the activation energy, which had its (quantified) beginnings in the Arrhenius equation mentioned previously. In modern chemical kinetics, the reaction path is represented in energy diagrams. These show in partial steps which amounts of energy are required to achieve a certain synthetic goal. For example, forming ammonia from the elements first requires breaking down the starting molecules. Under standard conditions the exothermic and exergonic reaction[29] is unfortunately very slow but increasing the temperature shifts the equilibrium situation unfavorably. It was discovered that significantly increasing the pressure improved the yield, particularly because this made it possible to achieve the high temperatures necessary for catalytic effectiveness. Without catalysts enormous amounts of activation energy would be necessary whereas with a catalyst (first developed primarily by Mittasch) only far smaller barriers have to be overcome (Figure 15).[30]

The first few of a whole series of adsorption processes on the surface of the catalyst are exothermic, and finally the ammonia molecule adsorbed onto the surface is released. The reaction components follow the path of least resistance (this correlates with Ostwald's "step rule"). As the formation enthalpy is a "state value," the energy balance is the same for the two variants "around the top" and "along the bottom."

Increasing the temperature cannot increase the reaction rate and thus the yield in every case and to any desired extent. In the best case heating leads to "autocatalysis" as increasing the total number of collisions also increases the number of effective collisions that result in the product. Yet it is important to understand the application framework. In his article *Über die Reichweite des Auslösungsbegriffes* of 1943, Mittasch states:

> In chemistry it is primarily the range of medium and low temperatures in which the actuation and directionality due to various influences play a major role. As chemical "impediments" continuously decrease with increasing temperature – the reaction rate increases greatly – one will hardly find any chemical actuation (of the usual sort) at 2000 degrees,

---

28 Steininger (2017, p. 28) states: "Wilhelm Ostwald's research on 'contact effects' from 1880 to 1900 tended to avoid solid objects."

29 The standard formation enthalpy is $\Delta H_B = -46$ kJ/Mol; the free standard formation enthalpy $\Delta G_B = -16$ kJ/Mol. This is thus a "voluntary" (spontaneous) process.

30 Cf. Ertl (1990) as well as Ertl & Soentgen (2015, pp. 161–168). Gerhard Ertl received the undivided Nobel Prize in chemistry in 2007.

but all the more so in the lower temperature ranges in which people normally observe and perform chemical reactions. (Mittasch 1948, p. 484)

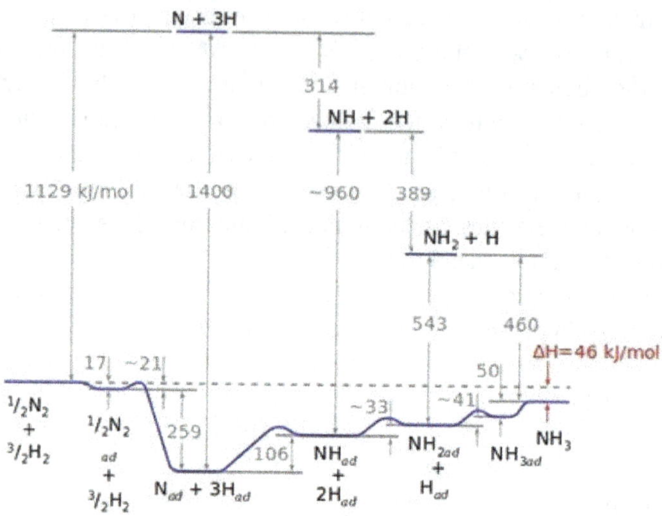

Fig. 15: Mechanism of the formation of ammonia by a catalysed reaction of nitrogen and hydrogen according to Ertl (1990).

This is a crucial specification of the catalysis concept and essentially of chemistry as a whole. At extremely high temperatures every substance disintegrates and if no more chemical reactions are possible then it is obviously pointless to talk of actuation and acceleration. Mittasch as well as Nobel Prize winners Fritz Haber (1868–1934) and Carl Bosch (1874–1940) could hardly imagine what exactly the reaction mechanism looked like and why their ammonia synthesis was successful. The mechanism was only deciphered with modern methods of surface examination (cf. Ertl 1990). In this case, too, the practical success preceded the scientific knowledge.

6.4 Conclusions

The questions of the possibility and reality of chemical reactions are for the most part addressed by studies of measurable heat phenomena. Upon closer inspection it becomes clear that no chemical reaction progresses to completion. This results in problems in determining the beginning and end of chemical reactions.

What appears macroscopically as a resting state may not be still in the submicroscopic realm. As a rule, such questions, as are typical in chemistry, are resolved pragmatically, that is by conventions. Ostwald assumes that a possible chemical reaction begins instantaneously when the reactants are present. In his view a catalyst accelerates a reaction but does not cause it. The notion of chemical potential developed historically from the theory of affinity. Modern chemistry is still essentially based on the idea of affinity although the concept has been obscured or superseded by attempts at quantification.

The field of catalysis poses other questions about the causation of chemical processes and thus about causality in chemistry. In any case, catalysis has a substance background whose interpretation began with the causality of contact. Although the Arrhenius equation represents a putative "energetic" approach in the treatment of temporal chemical questions, chemical thermodynamics continues to be the focal point for the examination of chemical potentiality. A coherent theoretical unification of potentiality and temporality is not in sight because the variable time does not exist in chemical thermodynamics.

Whereas the driving forces of chemical processes are dictated by natural laws and therefore only manipulatable within limits, the chemists, by creatively designing the reaction ("kinetic control"), have enormous influence on the reaction rate and the reaction path. Designing the reaction usually involves the skillful handling of kinetic impediments. Temporality in contrast to potentiality is thus not determined. The possibilities of designing the course of the reaction in particular are what makes chemistry a systematic art.

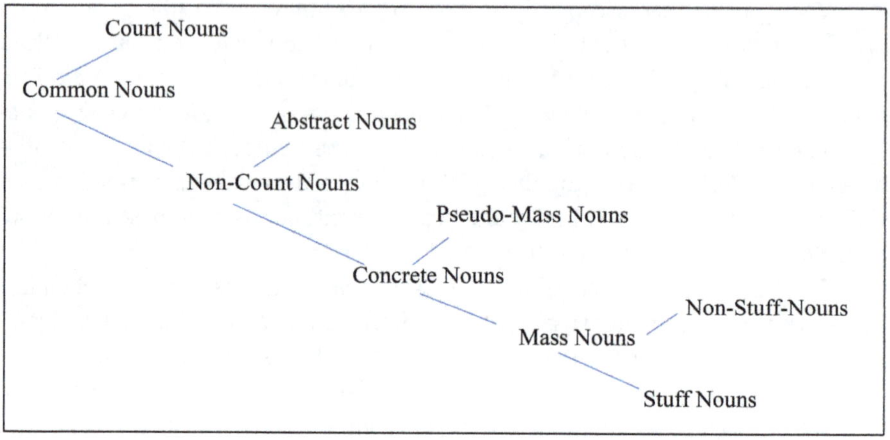

Fig. 16: Systematic grouping of nouns according to Peter Hacker (1979, p. 246). Substances fall under non-count nouns and exhibit the features consistency, dissectivity, and form independence.

7 Naturally Chemical?

To what extent can we call chemical substances natural? This chapter examines the so-called "natural" kinds in chemistry. We encounter substance-related classification concepts such as "element," "compound," "alcohol," and "acid" in the very first hours of a chemistry course. Chemical examples such as gold and water are used in almost every discussion of the term "natural kind" in the philosophy of science. These discussions often pose the question of the naturalness of the reference objects of this classification. As we will see, this question can be answered in completely different and occasionally contradictory ways.

In the first section I will attempt to clarify the concept of nature or, respectively, the concepts of nature with respect to the culturalist approach of Peter Janich. The second section is devoted to the question of what meanings the term "natural kind" can have. I will argue that referring to natural kinds introduces far more vagueness than clarity and the term should therefore either be clearly specified or not used.

7.1 Culturalist Insights

In an essay on the relationship between the expressions "natural" and "artificial," Peter Janich concludes that the dichotomy of these two terms does not itself belong to the linguistic inventory of the natural sciences but to the terminology of the history and philosophy of science, that is to say, it pertains to speaking *about* the natural sciences. He feels that the credo of modern physics, which most chemists regard as the leading discipline, is that everything (including speaking about nature) is natural. He calls this view naturalistic. Chemistry follows physics in the following sense:

> The argumentation paths in chemistry, too, lead *from the recognized theory to the perceptibility of its objects.* (Janich 1996a, p. 59; original author's emphasis)

Janich identifies three "great leaps" of chemistry on its way to the "naturalistic" modern period: the introduction of precise physical measurement technology, the organizing effect of the successful "atom-molecule concept" including the building-block metaphors it entails, and the anthropomorphizing of these building-block metaphors. In the course of these developmental leaps, chemistry, like the other natural sciences, has lost a clear concept of nature. Janich lists the following four component meanings of the distinction between natural and artificial in an initial systematic overview (Janich 1996a, pp. 64–68):

https://doi.org/10.1515/9783111401041-010

(i) *The natural and the artificial as objects whose origin or emergence is of interest.* In this (originally Aristotelian) approach what is at issue is the production or manufacture by humans of that which is artificial. Examples include artifacts like gold teeth that have organic, natural correlates. In this component meaning the following applies:

> Distinguishing natural and artificial things means emphasizing that aspect that *defines the product (its function) as the purpose of human manufacture.* The rest is secondary, i. e., a side effect or "natural." (Janich 1996a, p. 65; original author's emphasis)

(ii) Nature as the essence of a matter (the prevailing recognition of a situation).

> Whereas [...] the previous distinction of natural and artificial with things and events served to explain their origin and thus pertained to a descriptive aspect capable of truth, here we have to do with a normative or prescriptive aspect [...] (Insufficiently separating the descriptive and prescriptive use of natural and artificial proves to be one of the most important errors that burden the discussion of chemistry). (Janich 1996a, p. 66)

(iii) *That which is inaccessible [Das handelnd Unverfügbare].* According to Janich, the natural sciences are indecisive in this component aspect:

> On the one hand, scientific-technical knowledge can press nature into service to any desired end and have it at its disposal [...]. On the other hand, the natural laws in particular are regarded as technically inaccessible boundaries [...]. Natural sciences also know [...] a concept of nature where "nature" is what is inaccessible, yet for lack of methodological reflection often regard as particularly determined by natural law that which has a particularly cultural, namely human character of establishing or manufacturing: the principles of natural science. (Janich 1996a, p. 67)

(iv) Natural and artificial as a method distinction or terms of reflection. Janich explains this component meaning using an example:

> If when taking a walk in the woods we find a key and obviously regard it as a lost artificial product and not an organic object, then we are doing something different than when we reflect on whether atoms and molecules, volume, mass and pressure, chemical reactions and compounds, and other "objects" of chemistry or the natural sciences belong to nature or technology. All of the examples mentioned stand for something technical [...]. That which is technical is always that which is manufactured, that which is natural is that which is found. (Janich 1996a, pp. 69–70)

Janich thus indicates that a fundamental difference between the predicative component meaning (i) and the reflective component meaning (iv) exists for the concept pair natural and artificial.

> It is only mankind's technical and theoretical arrangements for example to establish measuring and experimental procedures and to subject natural things [...] to these procedures that provide the basis for our judgment that a body has a certain weight or a certain chemical composition. (Janich 1996a, p. 69)

Placing a found "organic" object in a substance category (for example mixture, pure substance, or element) requires undertaking certain systematic actions beforehand, that is technical manipulations with (or about) this object or representative parts of it. This important fact corresponds to František Wald's preparation doctrine which was discussed previously. Although the shorthand of the chemical specialist's daily routine might suggest otherwise, in reality we must invariably consider the required technical manipulations as well when we speak of chemical entities.

Janich is highly skeptical of a philosophical attitude that amounts to a purely theoretical conception of the world:

> It is a philosophically thoughtless invention to construct from the Aristotelian concept of nature as that which was not made by mankind the concept of nature without mankind. Such a construction would not be internally consistent and even unadulterated nature as a norm is always defined only relative to the omission of known actions because of undesired consequences. (Janich 1996a, p. 76)

The explanation in the final clause is directed against attitudes hostile to technology and science in the sense of "back to nature." As we will see, Janich's argument can also be turned against certain formal theoretical propositions of analytical philosophy.

What we want to regard as "nature" is obviously less the result of work in the natural sciences and far more dependent on the premises chosen for describing it. Janich's component meanings or modes of what is natural or artificial relate in different ways to the science of substances. That which is artificial in the Aristotelian sense of "handmade" (component meaning i) can lead in chemistry to the same product as something existing outside the laboratory. Water that appears in a water trap in a saponification experiment is – if one does not excessively limit the framework of the experiment – identical to the substance that precipitates onto the windshield of a car in winter.[1] An "organic" substance existing outside the instrumental world constructed by human beings, extracted, and characterized in some plausibly coherent manner cannot become another as a result of human intervention. In this sense, component meaning i is only conditionally applicable to chemistry, whereas it is *naturally* (component meaning ii) applicable to

1 Here Ostwald's *hylotropism* is appropriate.

artificial and natural teeth and to artificial and natural flowers. The expression "identical to natural" used in product advertising accordingly has a deeper metachemical meaning that is quite the opposite of the commercial intent: There is no distinction between the artificial and natural origin of the same substance in a product. The product is the same even when the path to its formation is different. Interestingly, in the discussion of natural kinds, hardly anyone seems to care about this aspect, nor do they care about the circumstance that all substances that humans process come directly or indirectly from nature.[2]

The reflective interpretation of that which is natural or artificial (component meaning iv) is particularly interesting for examining chemical phenomena. The necessity of a technical intervention to more precisely characterize chemical entities blurs the boundary between "natural" and "artificial" in no small measure. This is because a successful substance examination requires synthetic steps. Elements (in the modern sense) are not simply found in the local environment or in nature. They, too, must be prepared. The systematic art we call chemistry works at the interface between that which is artificial and that which is natural. Proceeding from this systematic outline of artificiality and naturalness, I will now move on to examine the conceptual framework of natural kinds in chemistry.

7.2 Natural Kind – a Popular but Ambivalent Term

For an initial orientation, I will begin this section with a few examples of definitions and will then examine the general concept of substance based on a study by Peter Hacker. My intention is to identify starting points for distinguishing between artificial and natural kinds. Continuing on, I will then discuss Jaap van Brakel's fundamentally pluralistic approach, the concept of the "chemical individual," and finally the capability of the causal theory of reference as applied to the acid-base theories.

7.2.1 Definitions

The British/Irish philosophers Alexander Bird and Emma Tobin characterize the concept *natural kind* as follows:

2 Janich also notes this fact (1996a, p. 65).

> To say that a kind is *natural* is to say that it corresponds to a grouping that reflects the structure of the natural world rather than the interests and actions of human beings. (Bird & Tobin 2024; original authors' emphasis)

Another definition of "natural kind" is found in the glossary of a text compilation on the philosophy of science:

> A type of property, process, state, event, or object studied by science, mentioned in scientific laws, and assumed to be a causal feature of the world. The primary instances of natural kinds are objects of scientific taxonomy, such as electrons in physics, zinc in chemistry, and species in biology. Natural kinds are contrasted with phenomena that are assigned no such systematic, organising role, such as an event's occurring after I drop this pen, or an object's being located 34 miles west of the Liberty Bell. (Boyd et al. 1991, pp. 778–779)

The German philosopher Jens Soentgen expresses it yet differently. In his general phenomenological substance classification system, he writes that substances can be natural kinds. He characterizes the latter as follows:

> I understand a natural kind to be such that is of itself a unit, whereas an artificial kind represents a unification, that is something that is brought together to form a unit by humans because of certain purposes and with the aid of certain methods. (Soentgen 1997, p. 101)

In the following sections I will refer back to the positions presented here.

7.2.2 Mereological Fundamentals

In a linguistic study of the substance concept, the British philosopher Peter Hacker arrives at an interesting manner of handling chemical kinds. He states:

> Substances, understood as material particulars, are the constituents of the world, the furniture of reality. Substance, in a different use of the term, is the stuff of which concrete individual things consist. Substances are made of, constituted of, some stuff or other, some substance or kinds of substance. It is in this sense of the term that we talk of chemical substances, or that we say of a thing that it is coated with a sticky substance. (Hacker 1979, p. 239)

Proceeding from "common nouns" and continuing to "non-count nouns," "concrete nouns," and "mass nouns" all the way to "stuff nouns," Hacker develops a systematic categorization of nouns, which he represents in a schematic diagram (see Figure 16). He concludes that the relationship between the conceptual structures of things[3] on the one hand and substances of which they consist on the other hand is complex, subtle, and therefore of great philosophical interest.[4] He refers to specific "non-count nouns" and "stuffs" ("The expression is ugly, but will have to do service") and identifies three characteristics for them:

> First, *consistency*, in the sense in which an item consists of something (stuff) or other. Second, *dissectivity*, in the sense in which a feature is dissective to the extent that if a thing has that feature, every part of the thing has that feature [...]. The third characteristic to invoke for present purposes is *form independence* and *specific quantity retention*. (Hacker 1979, pp. 242–243; original author's emphasis)

Composition (or the quality of being composed), dissectivity,[5] and form independence (with conservation of mass[6]) are for Hacker the prerequisites for the concept of substance. The latter is formed in dealing with things in our surroundings or used for that, but substances are not things but kinds, and: "A thing is never identical with the stuff that, at a given time, is constitutive of it." (Hacker 1979, p. 259) Hacker explicates the important qualification of substance dissectivity:

> [...] gold is dissective down to the atomic level, but beyond that point it is not, for an atom of gold consists of things – subatomic particles, not stuff – and those things do not consist of gold. (Hacker 1979, p. 256)

This "relative dissectivity" of substances is an empirical finding. This also means that the discussion of atoms and their characteristics on the one hand and the discussion of substances "consisting" of these atoms on the other hand belong to different categories and should be carefully differentiated. Hacker describes the situation as follows:

> So it emerges that things do indeed underlie stuff, and that the notion of S[tuff]-elements is ontologically more basic than that of stuff (although not epistemologically or conceptually so). (Hacker 1979, p. 256)

3 Hacker states: "[...] stuff [...] does not always constitute things." (Hacker 1979, p. 253). Gaseous and liquid entities facilitate a more immediate access to substances.

4 In contrast, this topic receives scant attention in the modern philosophy of chemistry and there has been hardly any response to Hacker's studies in this area. Exceptions include Jens Soentgen (2008) and Jaap van Brakel (2012).

5 Another name for this metachemical feature is "portionability," cf. Soentgen (1997).

6 I feel the latter features could also be listed separately. The classic ponderability does not appear in Hacker's writings as a separate concept.

The things (which, however, are not part of the world of everyday life) that in this regard underlie substances are, for example, the atoms in an elementary metal or the molecules in a compound. Disregarding for the moment that in experimental reality isolating these atoms and molecules is extremely difficult and that they are therefore not directly observable, it would be a mereological projection, if not a mereological fallacy,[7] to think that by describing their (indirectly inferred) features the true nature (or essence) of the substances whose constituents they are would be rendered understandable. Hacker does not use such designations in the study in question here but leaves no doubt that he himself is not inclined to essentialism:[8]

> To be sure, to understand a given stuff noun, to grasp the concept of stuff of a given kind, one must know the criteria of identification or application – the defining features that identify that kind of stuff (however flexible, open textured, or porous the concept may be). (Hacker 1979, p. 247)

Complete determination of its composition enhances our knowledge of a substance but it is not necessary to give it a name that will stick (its "baptism"). For example, the German name for nitric acid "Salpetersäure" [saltpeter acid] became the generally accepted designation from what was originally a wide variety of names;[9] it is a perfectly satisfactory name in that it describes the substance's manufacture, its descriptive features, and its chemical behavior. The analytic empirical formula (HNO_3) adds to this list of features (and may be ontologically more fundamental) yet from a methodological perspective it is a supplementary piece of information and one that in chemical practice is of secondary importance. Accordingly, we may conclude that in the present context as well, and in a manner particularly typical of chemistry, the empiricist positivist principle applies: the manifest image of the relationships in question has priority over the scientific image.

Except for the title and the introductory remarks quoted, Hacker does not discuss natural kinds in greater detail. Directly after his schematic diagram he provides a list of examples:

---

7 I have taken the expression "mereological projection" as "any inference from products [of experimental analysis] to constituents" here from Rom Harré (2014, p. 80). The classic case is "H_2O." The "mereological fallacy" was introduced by Bennett & Hacker (2003) within the scope of the philosophy of neurosciences. Harré applies it to matters in the philosophy of chemistry.

8 In his essay *Substance: Things and Stuffs* he is very clear about this (Hacker 2004, p. 48): "It is an illusion that scientific discovery can disclose what the words we use, such as 'gold' and 'water,' 'fish' and 'lily' really mean. For what a word means is determined by convention, not by discovery – although, of course, discovery may be elevated into convention by agreement on a new rule for the use of the word."

9 Cf. the appropriate articles in Ruthenberg (2020a) and the additional literature therein.

> Stuff nouns thus include names of natural elements (gold, uranium, chlorine), chemical compounds, physical admixtures, natural kinds of material, foodstuffs (wheat, barley), edible flesh (mutton, beef, venison), manufactured materials (rope, plastic), etc. (Hacker 1979, p. 247)

All species listed are assumed to be in "finished" form, and natural and artificial forms are not explicitly distinguished. The "natural elements" mentioned are listed here together with clearly synthetic products such as plastic, and "chemical compounds" can undoubtedly be both natural and artificial kinds. Elements and certain materials are "natural." A plausible interpretation of Hacker's conceptual analysis suggests that all substances fit into this classification system. The starting materials of *all* the entities in the quoted list are natural in the sense of Janich's aspects i or iii – subject to more extensive examinations.[10]

7.2.3 Chemical Species Viewed from a Pluralistic Standpoint

The Dutch philosopher Jaap van Brakel devotes a section of his monograph *Philosophy of Chemistry* to the critical discussion of "natural kinds" (van Brakel 2000, pp. 58–66).[11] In this text – which continues to be the standard reference on this topic in the philosophy of chemistry – van Brakel shows that a discernible consensus exists neither regarding the criteria for or definition of natural kinds nor regarding the naming of typical examples. Chemical substances such as water or table salt and elements such as gold or zinc are mentioned relatively often. Attempting to summarize, van Brakel names the following four aspects as crucial criteria: similarity/comparability, occurrence in natural laws with corresponding predictability of certain characteristics, origin, and nature/essence. To illustrate this, he refers to a biological species and states:

> First, we may draw on morphological or perhaps functional similarity. This is nearest to the approach of grouping those things together that resemble one another in manifest characteristics. The second possible criterion is that of interbreeding, which might be taken as a typical example of a projectible predicate involving laws or regularities of nature, without much theoretical embedment. Third, evolutionary theory in its neo-Darwinian form, has led to the descent criterion of species. Here, what matters is the origin of the entity. The fourth criterion, that of species as clusters of individuals bearing strong genetic similarities, is the archetype of a natural kind with an essence. Here what counts is underlying deep structure, not manifest

---

10 Not all "found" materials (see Janich's component meaning iv) are necessarily natural because anthropogenic technical substance transformation can bring artificial materials into the environment. The global problem of plastic in the oceans is not the first illustration of this fact.
11 Essentially van Brakel's entire book is a study about chemical kinds.

> similarities, macroscopic scientific projectibles, or common origin. (van Brakel 2000, pp. 59–60)

Van Brakel emphasizes that this is merely an illustration to provide a framework for the following discussion and not a finished, generally applicable methodology. Regarding similarity, he states that this is applicable at every reference level (manifest, scientific, metaphysical) and it cannot be used to establish a reliable distinction between natural kinds and any other kinds.[12] Van Brakel comments on the second aspect, the predictability of certain characteristics that are presumably natural, as follows:

> I can predict/project how, in given circumstances, the following entities will behave: water, cyclopropenylindene, oak, tiger, policeman, car, red paint, red-after-image, angry person, schizophrenic person, army, and so on. Of course, some predictions may seem more reliable than others, but none are either absolutely guaranteed or completely arbitrary. In all cases there are ceteris paribus considerations and other uncertainties. (van Brakel 2000, p. 63)

It thus becomes clear that even the second criterion does not provide a generally applicable explanation of "natural" in "natural kind." At this point in the discussion one could fall back on the classification system Janich outlined and first identify one's own (possibly unreflective) basic premises. The fact that this is often not done is one of the key findings of van Brakel's study of natural kinds:

> In fact, natural kinds are often introduced by a philosopher of science at a crucial point in order to justify a not further argued preference or admiration, to ground induction in objective similarities for example, to explain why nature is lawful, or to ground the reference of scientific terms as in the causal theory of reference. (van Brakel 2000, p. 64)

Van Brakel rejects essentialist explanations for natural kinds. On the one hand, he states that in the broad range of possible characterizations either many features come into question as essences (for example for "tiger" he lists its dynamic physique, its functional structure, its genetic structure, its ecological niche, or all of these together) or none; on the other hand, the eliminative reductionist search for the ultimate building blocks of the universe would also quasi obliterate chemistry as such:

---

12 Here, van Brakel's criticism is essentially directed against Quine. Quine emphasizes – in an explicitly naturalistic perspective – the significance of similarity for the determination of natural kinds, see Quine (1969, pp. 114–138).

> If one is drawn to microreductive essentialism, one ends up eliminating all substances, being left at best with "quarks", "superstrings", or whatever "final physics" settles upon and something nobody, not even a few hundred theoretical physicists, can really grasp. (van Brakel 2000, p. 66)

Hence, extreme forms of conceptual substance decomposition can lead to mereological misinterpretations if one undertakes to reduce substance characteristics to the characteristics of elementary particles (cf. Figure 3). Quine, whom van Brakel singles out for criticism, states for example:

> [...] that it is a mark of maturity of a branch of science that the notion of similarity or kind finally dissolves, so far as it is relevant to that branch of science. That is, it ultimately submits to analysis in the special terms of that branch of science and logic. (Quine 1969, p. 121)

Thus in the final state the ideal natural science – presumably theoretical physics – should evidence the uselessness of referring to (natural) kinds. Quine nonetheless concedes a somewhat broader perspective to chemistry further on in his influential essay *Natural Kinds:* Whereas the dream of ideal mathematical-logical physics has its metaphysical attractions, but

> [...] there would remain still need also to rationalize the similarity notion more locally and superficially, so as to capture only such similarity as is relevant to some special science. Our chemistry example[13] is already a case of this, since it stops short of full analysis into neutrons, electrons, and the other elementary particles. (Quine 1969, p. 136)

Although Quine, as van Brakel also expresses it, advocates an ontological reductionism according to which kinds must finally be replaced by exact regularities, he nonetheless acknowledges the actual differences in epistemic interest between physics and chemistry. The fact that he speaks of the local inclusion of similarities relevant to chemistry indicates an aspect that I feel is essential to the science of

---

13 With this example, Quine means his reference to "comparative similarities" between molecules. He seems to assume that chemistry is essentially compositionism. The crucial passage – extremely awkward and far from chemical reality and therefore even more significant – reads: " Then, in principle, we might get at the comparative similarity of objects a and b by considering how many pairs of matching molecules there are, one molecule from a and one from b each time, and how many unmatching pairs. The ratio gives even a theoretical measure of relative similarity, and thus abundantly explains what it is for a to be more similar to b than to c. Or we might prefer to complicate our definition by allowing also for degrees in the matching of molecules; molecules having almost equally many atoms, or having atoms whose atomic numbers or atomic weights are almost equal, could be reckoned as matching better than others. At any rate a lusty chemical similarity concept is assured." (Quine 1969, p. 135)

substances. Yet to what extent one can speak here of superficiality remains questionable in my view.

The majority of the other *essentialists* also have a view similar to Quine's with respect to natural kinds, above all Hilary Putnam (1926–2016). In the influential essay *The Meaning of Meaning* he summarizes his understanding of the formation of meaning of the concept of "water" in a table (Putnam 1975, p. 269) that he wants to be understood as a "vector," that is as a coherent sequence to be read from left to right concluding with the entry of the extension (the essential feature), which is independent of the person (Table 1).

Table 1: Putnam's "vector" to characterize water.

Syntactic Markers	Semantic Markers	Stereotype	Extension
mass noun, concrete;	natural kind; liquid;	colorless; transparent; tasteless; thirst-quenching; etc.	H_2O (give or take impurities)

Except for the last column this table may also contain additional entries. In light of the topic of this chapter, it is interesting to note that Putnam assigns water the classification "natural kind" without any further explanation. One may conclude from the context that he means the liquid substance found in our physical environment. Putnam interprets the feature "H_2O" as the conceptual benchmark (the inflexible designator) for the reference, that is the substance of water. Yet is it really true that the chemical sciences are intent on finding an ultimate feature? And does it make any sense to speak of the essence of a chemical entity?

First, we should note that the empirical formula H_2O is the abbreviated result of an examination of the composition of a sample. This datum is abbreviated (or idealized) because not insignificant aspects are ignored, such as the formation of byproducts from the breaking down analysis and the empirical fact that all chemical reactions are fundamentally incomplete. This is in any case not the molecular structure of water but the specification of its composition in shorthand. Immediate projection into the conceptual realm of submicroscopic physics is thus not correct. Referring to H_2O molecules is certainly helpful in many regards, for example when discussing bonding states, yet water is in reality not simply a mass of such particles.[14] Here, too, the mereologically relevant model of an object made

14 Paul Needham (2017, p. 128) provides another description of the same situation that relies more on logic: "'Is H_2O' cannot apply to a single, isolated water molecule because 'is water' applies to matter that freezes at 0 °C and boils at 100 °C, which it doesn't make sense to say of an isolated

of LEGO® blocks is illustrative (see Figure 3): Assuming the individual blocks are lying in a pile, it is only possible to construct a desired object when one knows exactly how the blocks should fit together. Yet even if that succeeds, no conclusions can be drawn about this object (the whole) solely from knowledge about the blocks (the parts).

In *The Meaning of Meaning* Putnam introduces an imaginary situation in which the substance known on Earth as water on a fictitious Twin Earth possesses the same "stereotypical" characteristics but a different submicroscopic composition, namely "XYZ." With this imaginary situation, he seeks to illustrate his (then current) view that the meanings of designations of natural kinds cannot be mentally constructed but are determined by scientific results ("meanings are not in the head"). This effectively declares the respective current submicroscopic physics the goal of all science, and for all the painstaking study in the direction of microstructuralism, we must ask how we can know at all that we are studying a natural kind. Similarly to Quine, all substance entities are regarded as natural because from the outset he proceeds from a naturalistic (or realistic) framework.

What is particularly irritating about Putnam's position is that no reference is made here to the reality of natural science. When the substance water is identified in a chemical laboratory, then as a rule one refers back to the "semantic" and "stereotypical" features, secondary features in Putnam's scheme, not the identification of microstructural aspects. Logically, van Brakel arrives at the following conclusion about the imaginary situation with the Twin Earth:

> Perhaps the most disturbing fact about this literature is the insistence of sticking with an example that simply makes no sense at all. (van Brakel 2005, p. 22)[15]

He had already taken a skeptical stance on the concept of natural kind earlier:

> I argue that we are confronted with a dilemma. Either we must favour a plurality of natural kinds, embedded in a variety of manifest, macro-, and microdiscourses. Or we must choose to

---

water molecule." Cf. my summarizing remarks in Ruthenberg (2012b) and the literature listed there.

15 Putnam purportedly did not know any better. In their correspondence van Brakel had explained his view to Putnam as he describes in a footnote: "I am grateful to Jaak [sic] van Brakel for notes about this [examining substance]. However, I do not see myself in a position to accept his own view according to which it is consistencies of the 'phenomenological' characteristics (such as melting point and freezing point) by which the reference of substance terms is determined." (Putnam 1999, pp. 72–73). Reference to the actual chemical analytical methodology such as Ostwald describes should be sufficient to contradict this view.

> eliminate them completely, at least those kinds ordinary people might know of. (van Brakel 2000, p. vi)[16]

Hasok Chang arrives at a similar conclusion. In explicit contrast to Bird & Tobin's definition quoted above, Chang emphasizes the human influence on the interpretation of natural phenomena:

> [...] it cannot be denied that nature as a possible object of our knowledge is inevitably framed by human perceptual and mental categories. (Chang 2015, p. 43)

His characterization of natural kinds – a "deliberately vague and multifaceted notion"–also has a culturalist note and reads as follows:

> I propose that we regard a classificatory concept as a natural kind if it is sufficiently effective in aiding inquiry in the natural sciences. (Chang 2015, p. 42)[17]

With this expressly broad characterization (which the author sees as naturalistic) it is also possible to understand the four elements of ancient times as natural kinds, with which Chang would presumably have only minor problems. The contemporary descriptions of these elements by the three main ancient authors can be summarized as shown in Table 2.[18]

Table 2: Main criteria of the ancient four-element concept in comparison.

	Water	Fire	Air	Earth
Aristotle	cold/moist	warm/dry	warm/moist	cold/dry
Plato	icosahedron liquid meltable	tetrahedron warm bright, flame	octahedron light, ether fog	cube solid, hard
Empedocles	Nestis rain, sea	Zeus (Hephaestus) sun, warmth	Aidoneus/Hera ether, sky	Hera/Aidoneus solid, durable bone

16 Eric Scerri (2020) reviews van Brakel on the one hand and criticizes the microstructuralism of Kripke and Putnam, yet at the same time emphasizes his support for Paneth's dualism (cf. Chapter 9). Headley's (2020) plausible approach from an intercultural perspective aligns with the critical view of Kripke and Putnam, as does mine, and emphasizes the significance of chemical practice.
17 A precursor of this view may be found with Ian Hacking: "[...] 'natural kinds' is a name for what helps people do better. Natural kinds, in short, seem important for *homo faber*." (Hacking 1991, p. 114; original author's emphasis). Cf. Boyd & Boyd (2019).
18 Cf. Ruthenberg (1992, 1993, 2005).

The concept introduced by Empedocles corresponds to Chang's suggestion: At least in the forms elaborated in greater detail by Plato and Aristotle, this concept remained for many centuries a great classificatory aid to the natural scientists studying substances. It should be noted that whereas the designation "water" for example comes from everyday experience, it does not refer solely to the material that Putnam characterizes with "stereotypical" properties. Plato for example classifies gold under the elementary category water because it is meltable, that is potentially convertible to liquid form. Although the descriptions of the classic four elements and their methodological functions differ greatly from those of the modern elements, it appears possible and even helpful to continue to regard them as natural kinds.[19] In any case, their abstract semantic kernels are still present in science in the form of aggregate states (earth/solid, water/liquid, air/gaseous) and the driving, generating, and transforming forces (fire).[20] If one does not want to completely do away with talk of natural kinds, then a pluralistic attitude is advisable. However, in such a pluralism it then becomes necessary to clearly differentiate the use of such concepts.

One of the examples Hasok Chang uses in the source quoted is acidity. A theoretical bifurcation[21] in the interpretation of acids and bases has existed rather precisely since 1923. On one side stands the *protonism* initiated by Svante Arrhenius and mainly developed further by the Dane Johannes Nicolaus Brønsted (1879–1947), that is the conviction that the characteristics typical of acids (in aqueous solutions, by far the most common case in chemical practice) are attributable to the hydrogen ions; on the other side is the *electronism* suggested by the American chemist Gilbert Newton Lewis (1875–1946), in which those substances are termed acids whose molecules exhibit a deficiency of electrons ("electron acceptors").[22]

Long before a serious and unifying theoretical examination of the phenomenon of acidity gained speed in the late 18th century, practitioners in workshops and laboratories knew species of substances that due to easily perceptible phenomena, above all their taste but also their ability to dissolve limestone with effervescence and the color changes of certain herbal solutions, soon came to be referred to as "acids." The first known acid, familiar since ancient times, was vinegar, and the so-called "mineral acids," which were obtained by subjecting minerals to extreme

19 Böhme & Böhme (2004) is a very comprehensive study of this topic.
20 The conceptual connection to the phase-centric notions of Wald and Ostwald is obvious.
21 I take the expression "bifurcation" from Ian Hacking (1983, S. 84–85). Cf. Ruthenberg (2024b).
22 Regarding the development of acid-base theories, see Lowry (1915), Walden (1929), Luder & Zuffanti (1946), Jensen (1978), Simon (1980), and Finston & Rychtman (1982).

heat in the presence of moisture (in the form of water or alcohol), came along later.[23]

Chang (2016) plausibly describes the development of the natural kind "acid" as a gradual process in which the sensory characterization mentioned and the observation of the substance's reactivity constitute the first stage, followed by the first attempts at identifying a general essence for the representatives of this substance group[24] and the emerging protonist doctrine, set in motion by Svante Arrhenius' theory of electrolytic dissociation in 1887. This concept underwent further generalizations a good quarter century later, particularly by Brønsted. Chang identifies Lewis' electron-pair theory as the last iteration to date.

The question is which of the two concepts is now the "right" one and whether both can be regarded as natural. Chang makes it clear that Lewis' theory is not a further development of the generalization of classic protonist notions. The latter contention is occasionally made and Lewis himself contributed to this misunderstanding by claiming to have arrived at (or to have been able to arrive at) his concept based on the usual sensory perceptions. In his efforts to bring acidity closer to modern chemistry, Lewis attempted the intellectual balancing act of bringing the experiential substance world into the realm of chemical non-observables. Even years later, he still criticizes classic protonism:

> [...] any similar valuable and instructive extension of the idea of acids has been prevented by what I am tempted to call the modern cult of the proton. (Lewis 1938, p. 297)

Today's experts continue to use the word "acid" for both interpretations, although the term "Lewis acid" (and correspondingly, "Lewis base") has become generally accepted for the electronic interpretation. In laboratory practice, "acid" is still understood as the protonist, that is classic kind. Hardly anyone now attempts to combine the two interpretations as Lewis did, although modern chemists are aware of the fact that reactivities and bonding relationships can be readily explained using the Lewis[25] electron-pair theory.

---

23 For the significance of everyday experience in characterizing acids and bases, see Ruthenberg & Chang (2017).

24 In the course of these experiments, Lavoisier postulated oxygen as an acidic essence, and a short time later Davy showed that hydrochloric acid does not contain any oxygen. The element with atomic number 8 nonetheless still bears the misleading name that originally meant "acid source."

25 Expressions such as "electrophilic" and "nucleophilic," which were introduced by Christopher Ingold (1893–1970), correlate with Lewis' thinking or resulted from it; cf. Ingold (1934).

Hacking's criticism of Putnam – in this case with respect to the acid-base theories – is also directed against Putnam's failure to consider current scientific knowledge:

> This example [Brønsted vs. Lewis] reminds us that the notion of meaning is ill-adapted to philosophy of science. We should worry about kinds of acids, not kinds of meaning. (Hacking 1983, p. 85)

Putnam mentions only the protonist version of the acid-base theories and overlooks essential aspects of the discussion in the philosophy of science that are appropriate today (and were in 1975).[26] However, he is correct in saying that when the current experts speak of acids, they are still speaking of the same class of substances of which their colleagues spoke in the 18th century, who at the time characterized "their" acids by sensory means. This assumes that we are speaking of Arrhenius/Brønsted acids. Whereas Lewis acids and bases also have a substance background, they cannot be isolated in pure form like hydrochloric acid, nitric acid, and citric acid. For example, there is also no case in which a substance once baptized as an acid later lost this designation for good reason. However, there are a number of examples of substances that are referred to as acids but do not exhibit the entire group of defining features (sour taste, color change with indicators, reaction with lime, decrease in the pH value, etc.). These include a few amino acids, barbituric acid, uric acid, and the nucleic acids. Descriptivists will find this fact unproblematic, yet whoever seeks a defining feature (of an essence) will fail in these cases. Indeed this is not unusual when characterizing substances. For example, characterizing something as "metal" means giving it a series of readily recognizable, easily measurable properties or dispositions (such as solid state, high density, malleability, luster, fusibility, conductivity, typical reactivity, etc.). There are exceptions: There are liquid metals, and some metallic products are not ductile due to their great brittleness. Thus the designation "metal" is not immediately dropped if only one of these features is absent; only when several of them are absent do we increasingly tend to avoid using the designation (cf. the example of silver in Ostwald's writings discussed above). A plausible "metallic" essence is absent here, too.[27]

Thus there is no answer to the question as to which of the competing acid-base concepts is the right one. The protonist version refers to different phenomena than

---

26 Cf. Putnam (1975, pp. 311–313).

27 With respect to the chemical species, it appears as if an essentialist description would be applicable only to the elements, if at all – assuming an ontological perspective. Cf. the discussion in Chapter 9.

the electronic version. The former is an approach developed for aqueous solutions, the latter an abstract theoretical image without a substance correlate and therefore without the possibility of quantification or measurement. The reason for the difference in these approaches lies in the simple fact that hydrated hydrogen ions are something with substance whereas electrons are not.

As the classic concept is closely linked to everyday experiences accessible to the senses, the designation of natural kind for the corresponding entities of protonism would probably be indisputable. Whether one would also refer to the Lewis acids and bases as natural kinds depends more strongly on one's fundamental attitude toward chemistry. Those who regard chemistry as a science of substances would view the Lewis acids and bases more as theoretical constructions and would be less inclined to use the expression natural kind; those who regard chemistry as the science concerned with the release, acceptance, or distribution of electrons between atoms or molecules would certainly have no problem with regarding these Lewis entities as natural kinds.

Fig. 17: The descriptivist (left) and the essentialist (right) reflecting on acidity.

Both versions correspond to Chang's liberal definition of natural kinds because both are useful in the scientific process and play important roles. If we apply this liberal definition we can also describe both as natural kinds, bearing in mind their difference:

> Lewis's concept of electron-receptivity designates a coherent and interesting property, but calling it "acidity" is a misnomer, just as "oxidation" is a misnomer to the extent that it doesn't have anything inherently to do with oxygen. (Chang 2015, p. 42)

As important yet long unrecognized the bifurcation he identified is, Hacking's criticism of Putnam is consequently unfounded because Lewis foresees totally different tasks for "his" acids and bases. In both "versions," "acid" and "base" mean something significantly different, and in spite of his selective approach – against his better knowledge – Putnam was not at all wrong to identify the Brønsted entities as "genuine" acids.

7.2.4 Chemical Individuals

An expression related to natural kind and one of crucial importance to the science of substances is that of the chemical individual:

> As experimental production and individuation is both a material and an epistemic activity, actions of identification and ordering of objects may be interwoven with experimental production and individuation. (Klein & Lefèvre 2007, p. 74)

This statement by two historians illustrates that the classification of chemical species is not solely a conceptual act and that it is possible to treat these species – for example the acids – as a group of individuals which means to differentiate them from other species.[28] The characterization of the chemical individual necessarily precedes the classification as a kind or species. For example in order to create the class (kind, species) of alcohols one must first have examined some of their representatives. Is it then correct to designate individual pure substances as a natural kind as many do with water, or might it be more appropriate to regard the plausibly characterized chemical entity as "natural" if one does not want to do without this adjective? The composition and characteristics of every individual

---

28 Lewowicz & Lombardi (2013) regard "substance" and "individual" as separate ontological (metaphysical) categories and emphasize that a substance ontology fits chemistry in contrast to an individual ontology which fits physics. I use the term "individual" as the other authors in the history and philosophy of chemistry do in the sense of the differentiation of individual substance entities. In this sense it is also possible to assign substances the designation "individual."

substance are undoubtedly "inaccessible" (Janich iii) and "of itself a unit" (Soentgen).

The Dutch historian of science Reijer Hooykaas (1906–1994) notes in an essay that the differentiation of the concepts of species and individual is important especially for chemistry. These come from the classificatory disciplines of natural history such as zoology, botany, and mineralogy. In a historical overview of the formation of the concept of the individual in chemistry, which he sees as beginning in the 18th century, he shows how the notion of homogeneity guided the observations of early chemists before all specific experience and that compositionism initially did not play a role:

> From the [modern] chemical point of view "composition" is one of the main characteristics of a substance. However, it is quite understandable that at first only a qualitative similarity of composition was asked for and that little attention was paid to equality of quantitative proportions. (Hooykaas 1958, pp. 307–322)

As temperature measurements improved, practitioners found that phase changes (for example melting and boiling points) correlated with characteristic constant transitional temperatures. Supplemented by these "physical" substance constants, it became increasingly possible to better characterize substance products of the various experiments. However, the individuation of a chemical entity was independent of the knowledge of its precise composition (Hooykaas 1958, p. 311). Hooykaas summarizes his overview of the development of knowledge from the 18th century to the end of the Long 19th century as follows (Hooykaas 1958, p. 313):

1. There are chemical individuals (homogeneous phases); their main characteristic is homogeneity. Mechanical mixtures are excluded.
2. The formation of some of them is accompanied by heat effect, and their properties are not found by simple addition of those of their components. Gas mixtures and solutions of gases in a liquid are excluded, but this is not the case with salt solutions, alloys, and mixed crystals.
3. Chemical individuals containing the same constituents are reckoned to belong to the same species.[29] Some species show fixed proportions; within other species a continuous series of proportions is possible. (Both are the result of "chemical forces").

29 This metachemical conclusion aligns with that of Hacker (1979, p. 239) quoted above.

4. Species with definite physical constants (boiling point, etc.) appear to have a definite composition. These are "chemical compounds" in the more restricted sense.[30]
5. From the standpoint of corpuscular theory molecules of the same species are equal; their constituents cannot vary their proportion continuously but only in a discontinuous way. Every species (chemical compound) has a definite composition. Different multiple proportions of the same constituents indicate different species.

Hooykaas discusses František Wald's early approach in his short essay *Was ist ein chemisches Individuum?* in which Wald suggested the following definition:

> Chemical individuals are phases that have originated in a phase system with at least one independent variation and have maintained an observably constant composition throughout all independent variations compatible with the phase system. (Wald 1899, p. 15)

Item 1 of the summary essentially addresses this suggestion. Ideally, a chemical individual according to Wald is a perceptible object. Hooykaas describes this similarly:

> [...] in Natural History the individual, a concrete thing, is always more positive[31] than the species, which bears a more abstract character. (Hooykaas 1958, p. 317)

Because, as we have seen, he categorically keeps the chemical non-observables out of his approach, Wald systematically neglects transitional states, metastable phenomena, and unstable and theoretical entities. Yet his credo, the preparation doctrine, applies to the non-ideal cases just as it does to the ideal cases: Manufacturing desired chemical products, states, and situations invariably involves manipulating substance entities.

Hooykaas correctly notes that Wald and Ostwald have not completely abandoned compositionism despite their criticism of atomism.[32] For Wald's "observably constant composition" he therefore suggests "constant properties" (Hooykaas 1958,

30 Here it is useful to take the plural "constants" literally in order to avoid possible confusion with azeotropes or eutectics: Observing only one of these fixed points does not mean that the substance in question is a pure chemical individual.
31 "Positive" is understood here to mean "perceptible with the unaided senses" as is usual in positivism, i. e., in contrast to "abstract."
32 In the chapter on Wald, I described how his theory of chemistry was related first and foremost to stoichiometric aspects. Kurnakov (1914) took up Wald's approach and distinguished individuals with constant composition and those with variable composition.

p. 314), which is probably also what Wald meant. A strange phenomenon and one typical for chemistry arises here, which I would like to briefly describe. Normal water comes in various different (hylotropic[33]) forms such as rain, snow, hail, surface water, seawater, steam, and also dew. When it is a matter of establishing which chemical entity, individual, or species is present in each case based on samples of these forms, then the sample must first be converted into a form suitable for analysis. As a rule, this will involve phase transitions. If one uses the term species for the substance of water in each of its different forms then one must understand that the analytical test of the appropriate samples in reality takes place in that form which one is currently examining: For example to prove that a piece of ice is in fact water, one converts it into liquid form, electrolyzes the sample, and compares the results with the expected "constituents," in this case hydrogen and oxygen. The evidence of constant composition (constant characteristics) that Wald refers to in his definition of the chemical individual thus requires phase transitions. Thus one must invariably take the possible phase systems, not any one phase, as the starting point if one wants to characterize an individual.

As has been mentioned, the essential principles of the classification of existing objects come from the typical disciplines of natural science. Hooykaas discusses the applicability of the theory of types used in biology:

> The comparison between the one genotype and the many phaenotypes in biology readily presents itself, but we should not forget that these many phaenotypes in biology belong to *the same* species. (Hooykaas 1958, p. 319; original author's emphasis)

It seems possible to grasp the different (tautomeric) forms of a chemical individual such as water as analogs to biological phenotypes.[34] However, the analogy remains incomplete because there are no populations of chemical individuals.

7.2.5 The Acids in the Causal Theory of Reference

As was mentioned further above, Ian Hacking criticizes the applicability of Putnam's notions from *The Meaning of Meaning* to actual sciences:

---

33 It should be noted that Wald was one of the few natural scientists to actually use his benefactor Ostwald's term in a few places, for example in Wald (1918, pp. 205–206).

34 The "genotype" of a chemical compound would then be the empirical formula, regarded by the essentialists as such.

> A disadvantage of Putnam's treatises is that he prefers fictions to facts [...]. Nonetheless, he has succeeded in liberating us from the pseudo-problem of the incommensurability of meaning. In order to name electrons, we do not need a theory of names [...]. The certainty that a clearly inappropriate theory is not the only possible one is sufficient for us. (Hacking 1996, p. 142)

Hacking concludes that the actual historical development of our notions of the electrons aligns with Putnam's version of the so-called "causal theory of reference" (CTR[35]). For previously theoretical entities that prove to be "real" (or "natural") such as neutrinos, electrons, ions, and radicals, the first baptism is a speculative act. In other cases, such as the heat substance, the magnetic monopole, or phlogiston, such success has remained elusive. For substance objects that exhibit some or all of the features that Hacker lists, the baptism refers to empirically easily accessible phenomena.

In their attempt to improve the application of CTR to natural kinds, Kyle Stanford and Philip Kitcher have concerned themselves with one class of such substance objects that has been mentioned repeatedly. They describe the "classic" protonist acid-base interpretation in the tradition of Arrhenius as deficient because of its limitation to aqueous matrices and the fact that substances exhibiting sour character but containing no hydrogen would be excluded. The authors pose two questions:

> First, if the Boylean definition, or some similar definition proffered by his successors, had been adequate, why strain to produce a definition in molecular terms? Second, it isn't the entire complex of Boylean properties that set the standard from the mid-seventeenth century to the present. Instead some of Boyle's descriptions, and some of the descriptions advanced by his successors, are rejected to yield a list of privileged phenomenological properties that can later be invoked as a standard for a good molecular definition. How do the ones that survive obtain this privilege? (Stanford & Kitcher 2000, p. 117)

The authors answer the first of these questions by answering the second one:

> As we have seen, a principal motivation for causal theories lies in the possibility of discovering that some members of a natural kind lack properties originally used in picking out the

35 Kyle Stanford explains what he sees as the crucial aspect of CTR as follows: "On a causal theory of reference, this act of 'baptism' fixes the reference of the term by annexing it to the underlying constitution of the actual sample, so when the members of this linguistic community use the term 'gold' they manage to actually refer to gold (that is, all the stuff with atomic number 79) even if important aspects (perhaps even all) of the *descriptions* they associate with the term or their further *beliefs* about gold turn out to be profoundly mistaken." (Stanford 2006, p. 147; original author's emphasis)

> kind, and just this happened in the history of chemistry with respect to acids. Boyle's original list was modified because chemists hoped to use the properties attributed to acids to point to a common inner constitution and found that some of the phenomenological features of acids were more useful in doing so than others. (Stanford & Kitcher 2000, pp. 117–118)

From a historical perspective a few corrections must be made here. First, we should note that baptisms ("baptizing," "dubbing," "christening") for the members of the group of the acids began many centuries before Boyle. Acetic acid and the mineral acids belonged, as acids, to the usual inventory of alchemist and iatrochemical laboratories and workshops.[36] The recognition and fixation of the results of these baptisms took some time, and it may well have happened that, as with modern entities that are initially speculative, the actual existence was occasionally regarded skeptically.[37] My crucial point of criticism is this: The descriptive framing of the acids (and bases) by Boyle and long before him is, like that of the metals, so elastic that the absence of one single feature has no effect. The theoretical interpretation invariably comes after the empirical findings; this is trivial, but causal theorists have been known to brush it aside. Thus the acids remained acids regardless of whether Boyle and Lemery advanced their Platonic corpuscular notions or Lavoisier his oxygen theory. Boyle's descriptive "original list" was not modified, rather only his theoretical speculations were rejected.

What required and continues to require adaptation to reality are the theories, not the empirical observations as Stanford & Kitcher suggest here. This example of CTR is thus useless.

It cannot be disputed that in forming a chemical kind the hope (or also the knowledge, depending on the level of development) lurks in the background that one or several features of an identical inner constitution are present. Yet it is equally indisputable that there are observable (empirically easily accessible or substance-related) features and non-observable (scarcely accessible or theoretical) features. For example, it is not obvious that acetone is a ketone from the typical smell (and typical substance constants) of this entity, which is liquid at room temperature. Assignment to this class of substances is a complex process yet it is aided neither by the belief in a rigid designator nor assignment to a "natural" kind. For

36 Cf. Libavius (1597). See also the historical discussion of the relationship of acids, bases, and salts – still without electronistic aspects – in Walden (1929, pp. 27–69).

37 Here, too, a comparison with Bruno Latour's description in *Pandora's Hope* suggests itself (Latour 1999, pp. 113–144): First the existence of lactic acid ferment was questionable, then laboratory experiments underpinned its cohesiveness, it received an action name, and finally a proper name as well, with which it received its "protective envelope."

chemists it is sufficient if the sample examined exhibits stable characteristics and reproducibly matches the standard samples they themselves have manufactured.

7.3 Conclusions

The uses of the term *natural kind* in chemistry cover a broad range. On the one hand, everything of substance appears natural because it comes, directly or indirectly, from the human environment: All materials required for substance characterizations come from there and that leads to the possibly sobering conclusion that absolutely all substances can be termed "natural." On the other hand, opportunities for further differentiation result from the scientific interaction with this human environment. If one wants to continue to use the term *natural kind*, this can lead to bifurcations such as in the case *protonistic* and *electronistic* acids and bases. The original view (Brønsted) and the result of bifurcation (Lewis) can even diverge so far from each other that the latter leaves the category of natural kind and its representatives should better be termed theoretical entities. It is trivial that conceptual and linguistic analytical studies can be of great help in the philosophy of chemistry (Janich, Hacker). Yet especially where the relationship to substance and scientific practice is neglected, they can also be misleading (Quine, Kripke, Putnam).

The dichotomy artificial/natural has a particular twist for the science of substances: A chemical entity that is found in "nature" does not necessarily have to be a natural substance. Carbon dioxide for example can be a product of human or animal metabolism but can also be produced by technical combustion processes or "artificially" fermented wine. And merely because we find our plastic softeners (and not only these) from the substance group of the phthalates in Greenland's ice sheet, nobody would think of referring to them as natural.

Thus one can question the added value of applying the concept of "natural kind" in the philosophy of chemistry for reasons that lie less in the term "kind" than in the artificial specification "natural." Anyway, if we want to continue to use the term "natural kind" then pluralism or plurality becomes unavoidable.

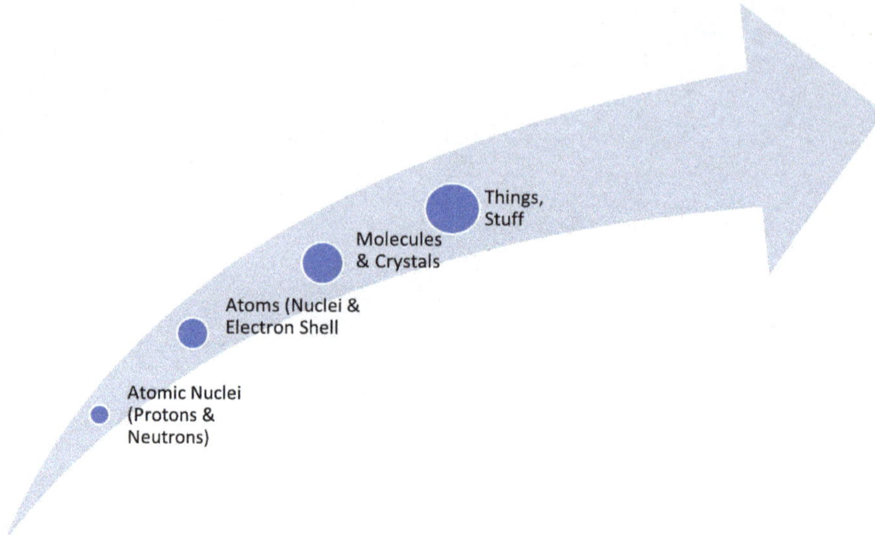

Fig. 18: Every range of possible epistemic objects in chemistry has its specific significance. Mereological projections are possible in both directions. Prioritizing or overemphasizing the significance of non-observable (theoretical) entities (nuclei, atoms, molecules) entails the risk of becoming excessively removed from the relation to real substances. Electrons alone are not part of the chemical realm.

8 How Chemical Is Quantum Chemistry?

The intrusion of theoretical deliberations and with them the discourse about unobservable theoretical entities into the realm of the science of material substances appears de facto to be unavoidable. This realm can be marked by the following question: What holds the constituents of a chemical compound together internally? An initial overview of the types of chemical bond can be obtained by studying the elements in the periodic system and their possible combinations. The majority of known elements have a metallic character; a smaller group consists of nonmetals. Three possibilities of combination result from this (from a modern standpoint): the metallic bond when metal atoms come together, the atomic or covalent bond when typical nonmetals combine, and the ionic bond when metals and nonmetals join.[1] This gives us a valuable yet rough classification system, but it provides no explanation for the bonds in the strict sense. Whereas a number of chemical lines of inquiry such as compositional aspects can be addressed descriptively and empirically, the theory of the chemical bond cannot entirely dispense with explanations of submicroscopic processes. This chapter will examine the application of findings of quantum physics to the chemical bond and its material background. In the second section, I will discuss the crucial direct precursor of modern quantum chemistry, the electronic theory developed and espoused by Gilbert Lewis in particular. The third section then goes on to examine the relevant works of a few theorists of chemistry, among them Linus Pauling and Hans Primas, and the conclusion summarizes the main points.

8.1 Electronism – the Precursor

Well into the 19th century, there were wild speculations as to which fundamental interactions could hold together the smallest particles of matter. One such speculation postulated an intrinsic "chemical form of motion" or a specifically chemical form of energy apart from very creative but empirically largely unsubstantiated ideas about atomism. Knowledge of the phenomenon which for over a century we have referred to as the "electron" marked a decisive advancement for theoretical chemistry.

1 This trichotomy is represented in modern textbooks, for example in the famous work by Holleman-Wiberg (1964), in which the ideal types mentioned are referred to as first-order bonds.

https://doi.org/10.1515/9783111401041-011

8.1.1 From the Theoretical Entity to the Electricity Atom – the Electron

As a *Rouse Ball professor of mathematics* in Oxford and the first professor of theoretical chemistry at that distinguished institution, Charles Alfred Coulson (1910–1974) writes in a survey article:

> The classical answer to the question: what is a chemical bond? is provided by the work of G.N. Lewis. It is the pairing together of two electrons, having opposed spins. (Coulson 1952, p. 1)

The prerequisite for the modern theory of chemical bonding is thus the conviction that the smallest building blocks of substances contain negatively charged elementary particles, electrons. Electrochemistry had shown that there was a close connection between the structure of chemical substances and electricity. Yet empirical evidence of the existence of electrons only came to light in the late 19th century. Prior to that, statements about the inner cohesion of elementary components in the compounds, and even in the metals, were purely speculative. Most of these statements were doomed to failure, some spectacularly so.

However, some speculations were to prove tenable and useful. One such successful assumption was expressed by the Irish physicist George Johnstone Stoney (1826–1911). He thought that electricity had a tiny elementary unit which he called the "electron." For adherents of the natural species, this is so to speak the baptism of the natural species *electron.*[2] This baptism itself has an interesting story. In a lecture in Belfast in 1874 (and when repeating this lecture in Dublin in 1881) Johnstone Stoney made the following statement:

> *For each chemical bond which is ruptured within an electrolyte a certain quantity of electricity traverses the electrolyte which is the same in all cases.* This definite quantity of electricity I shall call E_1. If we make this our unit quantity of electricity, we shall probably have made a very important step in our study of molecular phenomena (Johnstone Stoney 1881, p. 385; original author's emphasis)

This is a modern-sounding interpretation of Faraday's law applied to a theory of bonding. That law identifies a proportional relationship between the quantity of

2 Within the scope of his criticism of Kripke's and Putnam's theories of meaning, Hacking briefly outlines the early history of the electron and states: "No one at present doubts that the electron is a natural kind of fundamental importance." (Hacking 1983, p. 84) For the sake of neutrality, note that electrons in a culturalist perspective are theoretical entities or "constructs" (Psarros 1999, p. 306). Their "discovery" is thus essentially an invention (Hanekamp 1997, p. 227).

electricity produced and the quantity of material precipitated in electrolysis.[3] Here the decomposition aspect of the electrolytic experiment is conceptually reversed: If it is possible to break a chemical bond with electricity, then this suggests that the bond between the constituents is electrical in nature. However, in the quoted paper by Johnstone Stoney the baptism has not yet been completed. The author searches for the basic physical units (Johnstone Stoney 1881, pp. 382–383) and applies the very unusual terms "lengthine" (unit of length), "massine" (unit of mass), and "timine" (unit of time), as well as the "electrine" (unit of electricity), which is bonded to composite materials as described. He describes the latter in somewhat greater detail:

> e_1, the electromagnetic electrine, or the electromagnetic unit quantity of electricity in the metric series, is that quantity of each of the two kinds of electricity which must be discharged every second in opposite directions along a wire in order to maintain in it the metric unit current. (Johnstone Stoney 1881, p. 383)

It was only later that the smallest (negative) units of electricity were tentatively identified. In 1884 Johnstone Stoney felt obliged to defend his historical priority because a few colleagues referred to Hermann von Helmholtz, with respect to his Faraday lecture in 1881, as the first scholar to support the relationship discussed here.[4] Tellingly, Johnstone Stoney published this rectification under the title *Of the 'Electron' or Atom of Electricity.*

The introduction of the "electron" as an entity is a typical example of the *Latour iteration process* mentioned previously: The existence of many a theoretical entity is initially denied – yet once it has become apparent that there are good arguments for a "solid envelope," meaning that it is possible to work productively and consistently by assuming this entity, then it becomes part of the stock of science.[5] A good example of the skepticism about new entities is Ostwald's view of electrons. Even in the fifth edition of his *Grundriss der allgemeinen Chemie* he maintains:

> No experimental fact has yet made it necessary to assume a [...] granular or atomic nature of the ether, i.e. the space in which radiating energy acts. In contrast, an electromagnetic nature

3 According to Laidler (1993, p. 206), Faraday himself came close to a similar interpretation, although his notion of electricity as a fluid prevented him from coming to a strict atomistic and corpuscular interpretation.
4 Gilbert Lewis also failed to note the Irishman as late as 1923 when he wrote: "It was Helmholtz in his celebrated Faraday lecture of 1881 who first pointed out this deduction of the atom of electricity, or as it is now called, the electron." (Lewis 1923, p. 21)
5 Cf. Latour (1999, pp. 113–144)

> of the energy applied in radiation has recently become as good as certain, and for electrical energy, like mechanical energy, a granular or spatially inconstant nature has proven very probable. A large share of the electrical phenomena has been attributable to the action of electrical *elementary particles*, the *electrons.* (Ostwald 1917, pp. 554–555; original author's emphasis)

At this time, that is toward the end of the 19th century, physicists and physical chemists in particular had long since begun to correlate Joseph John Thomson's (1856–1940) empirical proof in 1897 of electrons in cathode rays, which Ostwald here implicitly refers to, with the chemical bond. Soon the very plausible opinion gained ground that these "electricity atoms" play a decisive role in the chemical bond.[6] Finally, in contrast to positively charged entities, they could be relatively easily "extracted," that is separated, from materials. Additionally, a detailed explanation for the electrolytic separation of substances was still lacking. However, Ostwald showed little interest in the theory of bonds even in the late phase of his scientific work.

Thus there were early forms of the electronic theory of bonds.[7] As early as 1908 Johannes Stark (1874–1957), who later won the Nobel Prize in physics,[8] published an article with the title *Zur Energetik und Chemie der Bandenspektra* in which he differentiates between "ring electrons" and "valence electrons" with respect to their functions in the atom. Ring electrons "constitute" the "positive electricity" whereas valence electrons "neutralize" the positive charge of the "electron ring" externally. Only the valence electrons are separable. Note that, as with Heisenberg[9] later, the term "valence electron" initially referred to the unbound free atom. Yet Stark also described electrons in a bonding function between atoms

6 An early indication of this is the statement by Ida Freund (1904, p. 540): "Not more than a beginning has been made as yet; but whilst formerly it had not even been possible to say what class of conceptions were likely to supply the building material for the adequate hypothesis and theory of valency that is required, it now seems probable that these conceptions will be the indivisible 'electron,' to which the ions owe their electrical charges and the lines of force connecting the electrons."
7 A detailed presentation of the history of this theory would go beyond the scope of the present chapter. I refer here to the exquisite standard work by Anthony Stranges (1982). Hanekamp (1997) provides a concise overview from a culturalist perspective.
8 He received the Nobel Prize in 1919 for the discovery of the optical Doppler effect in canal rays and the splitting of spectral lines in electric fields. It bears mention here that Stark later became a fervent supporter of National Socialism and the so-called "German physics" yet after the war went practically unpunished, see Mehrtens & Richter (1980) and Hoffmann (2005).
9 With respect to the discussion of the existence of stable atoms, Heisenberg speaks of binding energies and chemical forces; for him the binding of even one electron in an isolated atom represents a "chemical" bond (Heisenberg 1989, p. 89).

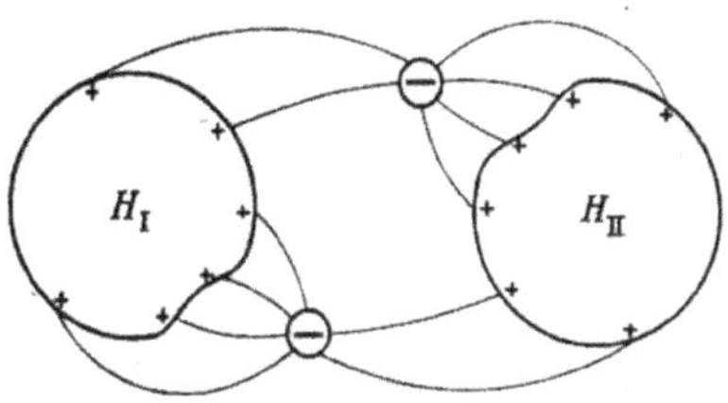

Fig. 19: Model of the hydrogen molecule according to Johannes Stark (1908, cf. p. 86).

as the illustration of his model of the hydrogen molecule given in Figure 19 shows.[10]

The idea is that the "force lines" of the "outer space" of one atom are strengthened by the approach of the "positive atom spheres" of the second atom with the result that a permanent bond is created. Stark now refers to the elementary particles involved as "saturated valence electrons." Although a genuinely chemical theory of bonds was not the objective in this case, we can assume that Stark's exposition at least promoted the formation of the electron-pair model of the chemical bond and maybe even gave rise to it. In his influential monograph *Valence,* Lewis refers to analogous illustrations from Stark (1915) of the carbon-carbon bond and the carbon-hydrogen bond when he states:

> In the last two illustrations we see the first suggestion of an idea which, as we are going to show in the next chapter, furnishes an extraordinarily simple explanation of the chemical bond, and completely reconciles the divergent views of the structural and dualistic theories. (Lewis 1923, p. 76)

The dualistic electrochemical ideas of Berzelius were revived as it were by the electron-pair theory while at the same time it now became possible to use the new theory to describe the occurrence of substances as varied as the inorganic salts (such as potassium bromide) and the organic compounds (such as methane). In potassium bromide for example, a bromine atom accepts an electron from a potassium atom and becomes a bromide ion while the body of the potassium atom remains behind as a cation. Both fulfill the octet rule. However, they do not form a molecule but an ionic lattice and remain as ions in the product. In methane, according to the electron-pair doctrine, four pairs of electrons form between the cen-

10 This is the reproduction of Fig. 41 from Stark (1915, p. 120). In the same place the physicist Stark speculates about the existence of the H_3 molecule. The author shows a figure with nearly identical content in the earlier work mentioned here: Stark (1908, p. 86).

tral carbon atom and the four peripheral hydrogen atoms. Here too, all the atoms involved have a complete outer electron shell, and they achieve the octet (carbon) or, respectively, a duet (hydrogen). However, the "simple explanation" provided by Lewis referred and refers to the electronic microstructure without answering the question of *why* the formation of electron pairs occurs. The standard answer to the latter heard in the lecture hall is: Because it is more favorable in terms of energy for the electrons involved. It is plausible that the unfavorable energy situation of mutual repulsion of the electrons is reduced when pairs of valence electrons are formed, thus increasing the mean distance between the electrons. Why it is precisely two electrons that form a bond became clearer if not any more vivid with the introduction of spin, which Lewis could not have known about in 1923.[11] The theory of the chemical bond based on electron pairs that developed in the decade following Thomson's experiments also accelerated the "triumph" of atomism in chemistry in the field of the conceptual interpretation of unobservable entities and processes. The extent to which atomistic concepts actually advanced chemical research, a primarily synthetic activity, is a separate topic.

The exceedingly fruitful thesis of Gilbert Lewis lacks a quantum mechanical background; in other words, his model also functions without a Schrödinger equation. The electron-pair model renders a number of important aspects of chemical practice accessible. Two of these aspects, which even today remain indispensable in the teaching and practice of chemistry, are discussed from an epistemological standpoint in the following section.

8.1.2 The Octet and the Oxidation Number

The beginnings of the octet concept, from which the so-called "octet rule" in the theory of bonds was derived, can be traced back to Mendeleev.[12] Still lacking a more specific notion of elementary particles, Mendeleev noted regularities based on his periodic system and sum formulas of stoichiometric compounds. These included the maximum valence of 8 in the row (period) of the following oxy-

---

11 A particle's intrinsic rotational impulse is referred to as spin in quantum mechanics. It can assume two values for one energy level and is thus a half-integer lacking a classical correlate and cannot be depicted. Only with the assumption of spin can all phenomena occurring in the line spectra be fully accounted for. The formation of an electron pair in a covalent chemical bond is not comprehensible in a classic sense because particles of the same charge should repel rather than attract each other. In the statement quoted at the beginning, Coulson projects the current state of knowledge a bit too far back.

12 Cf. here Russell (1971), Stranges (1982), and Jensen (1984).

gen compounds, whereby he assumed the bivalence of oxygen: K_2O, CaO, B_2O_3, SiO_2, P_2O_5, SO_3, Cl_2O_7, and OsO_4. Richard Abegg (1869–1910) also arrived at the number 8 in his theory of "electroaffinity." Abegg understood "electroaffinity" (not to be confused with directly measurable electron affinity) as a form of the tendency to ionization, that is a more or less pronounced inclination of atoms to form ions.[13] He defined "normal valences" and "contravalences" as follows:

> We shall describe in the following section those valences of each element which are of lesser number (<4) and therefore stronger as its *normal valences,* [and] those of greater number, weaker, of opposite polarity as its *contravalences.* (Abegg 1904, pp. 343–344; original author's emphasis)

Thus he assigns chlorine one negative normal valence and 7 positive contravalences and silver one positive normal valence and 7 (hypothetical) contravalences. With these basic assumptions, Table 3 can be constructed (Abegg 1904, p. 344).

Table 3: Chemical Valences according to Abegg (1904).

	Groups in the periodic system						
	1	**2**	**3**	**4**	**5**	**6**	**7**
Normal valences	1	2	3	± 4	-3	-2	-1
Contravalences	(-7)	(-6)	(-5)	± 4	5	6	7

Consequently, adding the valences together yields 8. This representation has had a lasting effect on the further development of the theory of chemical bonds. However, one should take into account that the initial concept of the electron was a different one than that which we have used since the introduction of wave mechanics in the natural sciences. For example, there was not yet any wave/particle duality and it was long thought that the counterpart of the negative electron was a positive electron.

Aside from the establishment of the octet concept, another precursor of a subsequent chemical concept may also be seen with Abegg. In direct connection with the table shown here, he discusses such things as the "polar character of an element in a given compound." He identified the following criteria for this (Abegg 1904, p. 344): ion formation, hydrolysis, its position in the periodic system with respect to its "compound partner," and the "formula of the compound in case the maximum number of the valences of the components come into play." At this point, the inclined contemporary reader will notice that electroaffinity exhibits

13 The first paper on this topic was the programmatic study by Abegg & Bodländer (1899).

a close relationship with modern electronegativity.[14] The latter, an approach to a quantified description of electrons in compounds, arose in the 1930s, initially with nearly no influence from quantum mechanics. The relational characteristic of the "polar character of an element in a given compound"[15] is estimated in modern chemistry via the difference of the electronegativities of the "compound partners." Abegg held the plausible view, one analogous to the application of electronegativity, that polarities occur especially when the respective electroaffinities differ significantly from each other. As an example, he cited the formation of covalent compounds between the elements in the upper range of a group in the periodic system (such as silicon carbide, SiC) and the lack of such compounds between the elements further below in the same group.

The octet rule, meaning the attempt of atoms and electrons to achieve a complete shell of eight electrons (in other words, the inert gas configuration), was first propagated by Gilbert Lewis. He depicted the electron configuration for the elements of the first "long" period in the graphically appealing manner that is illustrated in Figure 20.

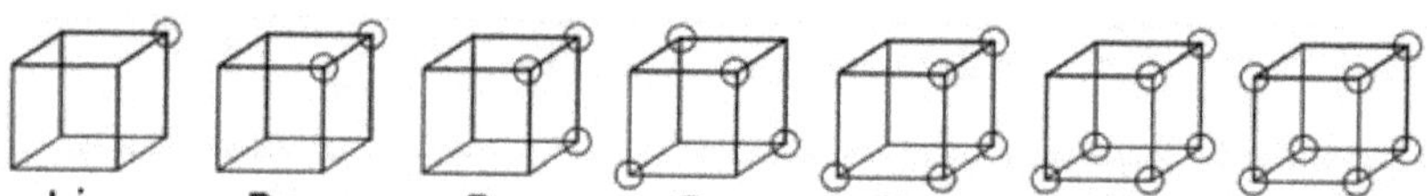

Figure 20: The cubic atom model according to Gilbert Newton Lewis (1916, p. 767).

Although the basic cubic principle[16] is purely speculative in nature, the relational inner structure of Lewis's idea is absolutely plausible. He enumerates a series of postulates behind this graphical representation, among them the following ones:

---

14 Something similar may be found in the short physiologically motivated work by Georg Buchner. The glossary entry under *electroaffinity* reads as follows: "The elementary atoms combine not only with other elementary atoms but also have the ability to bind electricity atoms (electrons) (electroaffinity). According to the unitary view of electricity, we refer to the elementary atoms connected with electrons as negative ions and the elementary atoms that have given off electrons as positive ions. The binding force for electrons increases with the *electronegative* character of the elementary atoms." (Buchner 1912, p. 136; my emphasis) Note that Buchner does not mention Abegg.

15 At this point I will skip over the classic problem of the preservation of elements when forming and separating composite substances and its semantic peculiarities and refer to reader to the respective discussions in other places (see Chapter 9).

16 Lewis, in the same place, calls the approach "the theory of the cubical atom" and dates its inception to the preparation for an introductory lecture on 28 March 1902. He notes that it was not the priorities that were important to him but the common ground with similar theoretical at-

> [2.] The atom is composed of the kernel and an outer atom or shell, which, in the case of the neutral atom, contains negative electrons equal in number to the excess of positive charges of the kernel, but the number of electrons in the shell may vary during chemical change between 0 and 8.
>
> [3.] The atom tends to hold an even number of electrons in the shell, and especially to hold eight electrons which are normally arranged at the corners of a cube. (Lewis 1916, p. 768)

Lewis emphasizes that the upper limit mentioned here cannot be exceeded. Since Irving Langmuir's (1881–1957) comprehensive "extension" of Lewis's ideas, this "rule of eight" has been referred to by the scientific community as the octet rule.[17] Although it plays no role in modern theoretical chemistry, which is essentially quantum chemistry, this rule remains an integral part of modern general chemistry. For example, the construction of correct valence structural formulas, that is electron distribution formulas, which Lewis pioneered, is unthinkable without the octet rule. Simple criteria are used in this construction: First, one needs the basic spatial structure, that is the arrangement of the existing constituents with respect to one another. This is easy for the diatomic oxygen molecule and accordingly more difficult for larger structures. Second, one needs the total number of outer electrons in the molecule (in the example $2 \times 6 = 12$). The electrons in the inner shell are not taken into account, as is also the case in quantum chemistry. Third, the octet rule comes into play in a heuristic sense: One now attempts to arrange these electrons symbolically in the postulated structure (in this case a linear diatomic structure) in such a manner that each individual atom symbol other than hydrogen receives 8 electrons. It is entirely possible that the number 8 is not reached even in existing entities. However, in this concept it may never be exceeded. The octet rule essentially functions as a law for the elements of the second period. Individual electrons are represented by dots and electron pairs by lines (according to the current convention). Mixed forms are also used in the literature. A

---

tempts (aside from Abegg, whom he mentions explicitly, certainly the one by Kossel as well), whereby the fact that "that all possess some characteristics of fundamental reality" would contribute to their probability. He does not have a realistic microstructural interpretation in mind.

17 Langmuir's introductory statement in this work is telling: "The problem of the structure of atoms has been attacked mainly by physicists who have given little consideration to the chemical properties which must ultimately be explained by a theory of atomic structure. The vast store of knowledge of chemical properties and relationships, such as is summarized by the Periodic Table, should serve as a better foundation for a theory of atomic structure than the relatively meager experimental data along purely physical lines." (Langmuir 1919, p. 868) The American industrial chemist Langmuir (who received his doctorate in 1909 in Göttingen) received the Nobel Prize in 1932 for work in the field of surface chemistry.

model of the O_2 molecule according to these rules can appear as shown (cf. Lewis 1923, p. 88; in contrast to the original, the bond electrons are represented as lines):

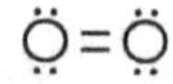

In this valence structure model, both oxygen atoms each have 8 electrons in their immediate vicinity and are connected by a double bond. The octet rule is fulfilled for both bonded partners. The chemist would say that these atoms have been stabilized by achieving the inert gas configuration (of neon). Nonetheless, this attempt cannot be regarded as completely satisfactory. Whereas diatomic oxygen exhibits what in chemistry is known as “double bond character,” meaning a specific behavior in reactions, it simultaneously has the character of a radical, which has been demonstrated empirically by its paramagnetic behavior. This radical character, that is the presence of unpaired electrons, is not expressed in this formula model. It was almost unavoidable that Lewis, the proponent of the electron pair bond, creator of the valence electron notation system, and discoverer of the relationship between paramagnetism and unpaired electrons in radicals (Lewis referred to “odd molecules”), would encounter this very same problem. In addition to the variant noted above, his notation also includes the following (Lewis 1923, p. 89):

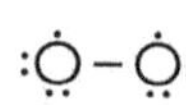

This model of the electron configuration of the oxygen molecule lacks the aspect of the double bond, and both oxygen atoms also fail to achieve the octet (they each have 7 electrons in their immediate vicinity). Lewis is well aware of the problems of a realistic interpretation of this dilemma:

> It must not be supposed, however, that such formulae are anything else than gross exaggerations of the state of affairs which is associated with the double bond. Perhaps they represent extreme states which [may] occasionally be attained by a few molecules. (Lewis 1923, p. 89)

Lewis does not offer any possible resolution to the dilemma. The modern application of Lewis structures uses the concept of “resonance” that Pauling propagated. Probably the best known example of this idea is the benzene molecule (C_6H_6): The three double bonds assumed to be present in this ring-shaped formation are delocalized, that is they cannot be associated with specific C-C groups. For this reason there can be no unequivocal electron distribution formula. The technical jargon occasionally uses the term “ring flow” of the electrons to characterize an “aromatic” structure such as this. If one wants to use the Lewis structure to represent the relationship, one will have to make do with two equivalent formulas shown side by side.

In reality, the electrons in the benzene ring are associated with one single state, albeit one that Lewis's model is unable to depict. The electrons do not "jump back and forth" between two specific structures. An analogous application would also make use of such a "resonance" double arrow in the two formula diagrams representing the oxygen molecule. "Resonance" or "mesomerism" refers to the fact that the same molecular entity can or must be represented with different electron distributions.

The relationship addressed here is one of genuine interest to the philosophy of chemistry because it poses the classic question of whether to argue for a realistic or an instrumentalist interpretation. The fact that the actual electron distribution is thought to lie "between" the resonant boundary structures can lead to misunderstandings, especially among beginners. Like all other chemical formulas, Lewis structures have the character of models and cannot be interpreted directly as realistic representations. Yet this is precisely what happened in the early 1950s when Soviet scientists rejected Linus Pauling's resonance idea, ostensibly because of the argument that the bond state had to be a precisely determined state. In reality, ideological and nationalistic tendencies played a larger role.[18]

In the everyday practice of chemistry, where theoretical considerations are of secondary importance, resonance has become accepted as a helpful concept and it is utilized in many applications. However, when modern chemists trained in quantum chemistry take a closer look they tend to be cautious about assuming its existence.[19] Indeed, they understand that neither classical concepts nor Lewis's abstracting formulas can adequately describe the unambiguous electron bonding state. If it seems useful to undertake a philosophical classification of resonance/mesomerism, then it would be best placed in the nonrealistic (instrumentalist) category.

---

18 Cf. the detailed discussion of this episode from the perspective of the philosophy of chemistry in van Brakel (2000, pp. 27–34). The author comes to the following evaluation: "The debates about the theory of resonance show that nationalistic concerns were of much greater importance than the dogma's [sic] of dialectical materialism. The concerns of the debates and power machinations in Moscow promoting Buterlov had no followers in the GDR – though it is no doubt correct that Butlerov's importance had been suppressed in German and English text books that gave reviews of the history of organic chemistry." (van Brakel 2000, p. 34)

19 Hans Vermeeren (1986) concurs. Valeria Mosini (2000, p. 578), however, presents an opposing viewpoint: "[…] as a matter of fact, the interpretation of resonance given by the chemists is a realist one, as the evidence from the history of the theory strongly suggests."

Such an interpretation may be found in the very enlightening discussion in Laitko & Sprung (1970, pp. 99–109), of course without the concepts used here. Regarding the function of the valence structural formulas, the authors state:

> Because the formula diagram largely equalizes the inter-atomic connections, it was to be expected from the outset that not every bond can be expressed sufficiently precisely by *one* formula using the means of the valence scheme. (Laitko & Sprung 1970, p. 99; original author's emphasis)

This is a variant specific to the philosophy of chemistry of the epistemological fact that models cannot be simple unambiguous images of reality. Rather they provide opportunities for discussing theoretical entities from different perspectives. It then becomes particularly interesting to compare these different perspectives with one another, as in this case the electron-pair concept in the valence theory with the molecular orbital theory (see below). The authors discuss how symbols and concepts are generally handled when "new areas of reality" are opened up. They identify two possibilities to do the latter: Either the concepts are further differentiated and new symbols are added (as in the Chinese language) or more complicated combinations of the same basic symbols are introduced. Laitko & Sprung make it clear that the second variant predominates in chemistry and cite the resonance notation (see the benzene example) as proof.[20] They also emphasize that these early approaches to describing the electron configuration have no basis in quantum mechanics.

The current discussion of tautomerism is closely related to the claims that resonant boundary structures exist. In contemporary chemistry, tautomers are defined as isomers that are in actual active chemical equilibrium.[21] Probably the best known example is the "keto-enol tautomerism," for example that of acetone:

Although this important solvent is commercially available as "dimethyl ketone" or "propanone" in "pure" form (that is as a chemical individual), it is not actually pure. Part of the product, albeit a small part, consists not of ketone (left) but of enol (right). The attempt to separate the ketone from the enol is as doomed to failure as the attempt to obtain pure enol. This chemical substance is not a static mix-

20 Cf. Ingold (1934).
21 Cf. Staab (1959, pp. 642–650).

ture. The identity of a substance in such cases assumes a non-logical form so to speak and in addition to this the situation described by Laitko & Sprung arises:

> The predominance of traditional structural thinking in the valence scheme led to the widespread view that all cases in which the usual structural formulas are inappropriate are in reality examples of tautomerism, and the inability to isolate the various forms with unambiguous valence structures is due solely to their excessively high speed of transformation. (Laitko & Sprung 1970, p. 100)

The differentiation between tautomeric and mesomeric phenomena (that is shifts of atoms and electrons) soon resolved several specious problems for the proponents of the science of material substances. Incidentally, water is also a tautomeric substance system[22] (which would also resolve a few specious philosophical problems).

A further pre-quantum mechanical concept of electronic chemical theory stems from the reasoning that the differing character of the bonded elements must influence the polarization of the bond: the oxidation number. This number is regarded as the virtual charge that a type of atom would assume if the chemical entity in which it occurs were to consist entirely of ions.[23] In the elementary state, regardless of the current allotropic form, atoms have the oxidation number 0. In salt-like compounds the assignment is easy: The charges determined are identical to the oxidation number, for example in LiF (Li^+, oxidation number + I, and F^-, oxidation number −I, are present together in this material, at room temperature in an ionic lattice). Additional rules are required for covalent compounds. For example in calcium sulfate ($CaSO_4$), the state of the calcium atoms is quickly determined with the oxidation number +II. However, a logical manner must be found to distribute the double negative charge of the sulfate ion over the five atoms present. This is done by assigning the oxygen atoms, which have higher electronegativity (see below), the oxidation number −II. This means that each oxygen atom can nominally “take on” two electrons from atoms of greater eletropositivity (from those of the sulfur) to achieve the octet. Adding all the oxidation levels of the oxygen atoms together yields the virtual charge of −8. Because the actual charge of the sulfate ion (resulting from the two electrons of the calcium atoms) is known, the sulfur atom can be assigned the oxidation number

---

22 Jaap van Brakel (2012) refers to this point.

23 Psarros (1999, p. 169) says of the oxidation number: “It specifies which valence level the atoms of a given nonmetal assume when they are combined with atoms of other elements.” First, this characterization is incomplete because it mentions only nonmetals and, second, it is unclear because the oxidation number has nothing to do with valence (itself a concept requiring clarification) but with the ratio of the sorts of atoms involved to the electrons available for binding (valence electrons), cf. Riedel (2004, p. 128).

+6, which is interpreted to mean that it now lacks all six outer electrons of the ground state.[24] This information has practical relevance when comparing the state of this sulfur atom to those of other sulfur atoms, for example those in the sulfite ion (SO_3^{2-}). The sulfur in the latter has the oxidation number +4 and is thus less strongly oxidized than in sulfate. That means, for example, that an oxidation must be performed to obtain sulfate from sulfite. In the early electronist phase without quantum mechanical assumptions,[25] results were obtained that have gone on to become an integral part of general chemistry. One of these results is the theory of oxidation and reduction discussed here, which Gilbert Lewis, shortly before the beginnings of quantum mechanics, refers to metaphorically in the following words:

> It is only in the last few years that we have realized that every process we call reduction or oxidation is the gain or loss of an almost imponderable substance, which we do not call phlogiston but electrons. (Lewis 1926, p. 168)

Here Lewis commits, on purpose, a categorical error – a mereological fallacy – because, like phlogiston, electrons are not material (not a substance).[26] Nonetheless, his analogy expresses the essence of the phenomenon. One may often search in vain for discussions of the octet theorem and even of the oxidation concept in modern writings in theoretical chemistry. Yet both are exceedingly valuable and typical achievements of chemistry.[27]

8.2 A New Mechanics in Chemistry

Quantum theory was established by Max Planck. This fundamental scientific achievement made it clear that not only does the material substrate exhibit a lower

---

24 With this deduction it is assumed (from experience) that the basic arrangement of the sulfate ion is such that the sulfur atom is in the center and the oxygen atoms arrange themselves around it. Compounds with oxygen atoms can be excluded here.

25 Naturally, the assumption that atoms and electrons exist itself presupposes that the samples of substances can be divided or quantized at the submicroscopic level. Yet quantum mechanics was not yet used here.

26 Nevertheless, this fallacy is customary in chemistry, see Dainton (1975).

27 There are arguments in favor of continuing to classify the octet rule as not merely heuristic but to take it seriously even from a theoretical standpoint. See Weidenbruch (1994, p. 20): "Lewis's octet rule has lost none of its validity since it was first formulated over ninety years ago. It applies to compounds of the main group of elements with the correct number of electrons, which include nearly all organic compounds, as well as to compounds with excess electrons in which the supernumerary electrons can be accommodated in nonbinding orbitals of the substitutes."

boundary, but energy does as well. In reality there is no continuum if one looks closely enough. Quantum mechanics, which blossomed in the 1920s, provided the foundation for the next important steps in the theory of the chemical bond. Although it occasionally seems to be a cure-all in chemistry, quantum chemistry can do many things but it also has its limits.

8.2.1 The Beginnings

Quantum chemistry is understood to be the application of quantum mechanics to certain chemical problems.

> Atoms and molecules are mechanical systems that cannot be accurately described by the classic dynamics of Newton or by that of Lagrange or Hamilton. Therefore, chemistry uses a new mechanics, quantum mechanics, for molecular problems. (Primas & Müller-Herold 1990, p. 22)

With this statement by chemistry theoretician Hans Primas (1928–2014) and his coworker Ulrich Müller-Herold in a textbook of quantum chemistry, it becomes clear ("molecular problems") that the chemical bond is the problem to which quantum mechanics is applied. As has been stated, many early nuclear physicists understood the bond between the electron and the nucleus to be "chemical." This internal atomic cohesion and the actual chemical bond, that is the more or less permanent connection of atoms to one another, cannot be consistently treated using classical concepts. First, electrons repel each other and, second, they should be expected to crash into the nucleus because they are drawn toward it. Yet because atoms in their basic state emit no radiation, a theoretical description of them requires additional assumptions. The traditional way out of this theoretical dilemma is the idea of emphasizing the wave character of electrons. The Schrödinger equation is put to use for this purpose (cf. Heitler 1961, p. 19): $\nabla^2\psi + 2\,m/\hbar^2\,(E\text{-}V)\psi = 0$.[28] This equation is a second-order partial differential equation for the spatial coordinates and a first-order equation in time.

Spectroscopic measurements provided the empirical basis for the development of quantum mechanics.

28 In this "wave equation" E and V are, respectively, the kinetic and potential energy of an electron in an electric field (between nucleus and electron), m is the mass of the electron, ħ in the Planck constant $h/2\pi$, and ∇ (the "Laplace operator") has the form $\delta^2/\delta x^2 + \delta^2/\delta y^2 + \delta^2/\delta z^2$ for the usual three-dimensional case. Where the mass of the electron is regarded as unchanging, we refer to the time-independent or nonrelativistic Schrödinger equation.

> Extensive experimental material, most of it obtained spectroscopically, shows that an atom cannot exist in states with continuously variable energy but only in certain discrete energy states referred to as the [...] "discrete energy levels." (Heitler 1961, p. 1)

The quantization of energy in radiation processes has become increasingly apparent since Johann Jakob Balmer's examinations of the solar spectrum in 1885.[29]

Nuclear physicists questioned why, for example, hydrogen atoms should form diatomic molecules. In thermodynamic terms, the fact of the formation of this molecule is described by the reaction enthalpy: For the reaction $H + H = H_2$, is $\Delta H = -436$ kJ/mol, meaning the process is exothermic. Such results would probably have satisfied the earlier (non-atomistic) energeticists like Ostwald. For the modern energeticists,[30] the question of microphysical relationships is key. The classic attempt to explain that the electron pair somehow shields the two nuclei from each other and realizes a gain in energy did not really convince them. They also failed to find an explanation in Gilbert Lewis's electron-pair theory.

Walter Heitler (1904–1981) and Fritz London (1900–1954) are generally named as the authors whose work initiated quantum chemistry in the strict sense (Heitler & London 1927). In this work, using simplifications and approximations (the only analytical solution of the Schrödinger equation for a composite particle remains that for the hydrogen ion H_2^+ with only one electron[31]), the authors conclude that the chemical bond between neutral atoms is explainable by quantum physics. A few decades later Heitler expresses this conviction, which has since become generally accepted in the scientific community, as follows:

> *The reason for the formation of the molecule is the quantum mechanical phenomenon of electron exchange* and the main share of the binding energy is provided by the exchange energy A. (Heitler 1961, p. 99; original author's emphasis)[32]

However, the simplifications undertaken cause calculations to deviate from proven experimental findings. When the process is reversed, the reaction enthalpy mentioned above has a positive sign as dissociation energy and is a measure of the

---

29 The wave number of the "Balmer series" is expressed today as $\tilde{\nu} = R_\infty (1/2^2 - 1/n^2)$, with the Rydberg constant R_∞ and n = 3, 4,...

30 I refer to the modern theoretical physicists and chemists in this manner because the Schrödinger equation at its core has to do with nothing else than potential and kinetic energy.

31 Cf. Ladik 1973, pp. 203–212.

32 Regarding "electron exchange" Heitler states: "The electron exchange is [...] primarily a typical quantum mechanical effect and all attempts to find a 'classic analogy' must fail [...]. The cause of the exchange effect is the very fact that electrons are indistinguishable and therefore an exchange of electrons is principally unobservable." (Heitler 1961, pp. 99–100)

strength of the bond in the molecule. Only about two-thirds of this enthalpy is achieved in Heitler & London's calculation, and the internuclear distance that they calculate is greater than the observed value (cf. Pauling 1962, p. 23, and Figure 21). Heitler & London's approach can thus explain covalent bonding in qualitative terms, but in quantitative terms it does not lead to a fully valid theory in the usual sense, only to an approximation (as was to be expected). This is expressed by Pauling as well. He writes:

> The energy of the electron pair bond is for the most part resonance energy that is attributable to the exchange of the two electrons between the atomic orbitals. (Pauling 1962, p. 24)

This is the description of the situation obtained by applying the resonance concept. It appears to come very close to Coulomb's "classic" explanation, yet we may not ignore the fact that resonance is an unobservable (unmeasurable) thought construct. János Ladik offers a surprisingly vivid, indeed classic explanation for the covalent bond:

> In summary, based on the Heitler-London calculation, it can be said that the covalent bond between two atoms occurs because the internuclear distance has a value such that the negative energy of the interaction of the electrons with the two nuclei has a larger absolute value than the sum of the positive energy elements, which result from the potential energy of the repulsion between the two nuclei and the two electrons as well as from the increase in the kinetic energy of the electrons. (Ladik 1973, p. 217)

This explanation is based on Coulomb's law of electrostatics and, in contrast to Heitler's interpretation, quantum mechanical principles are rather secondary. The potential curves as a function of the internuclear distance of the two hydrogen atoms can be represented as follows:[33]

Heitler discusses another aspect related to chemistry which he refers to as "saturation":

> On the basis of classical physics it is absolutely impossible to understand why two neutral atoms, such as two hydrogen atoms, form a molecule. Additionally, the covalent bond possesses the extraordinary characteristic of exhibiting *saturation* as is expressed in the concept of *valence*. A hydrogen atom is able bind another hydrogen atom but not two or three. A carbon atom can bind four hydrogen atoms but no more [...]. A phenomenon such as saturation is completely alien to classic physics. (Heitler 1961, p. 91; original author's emphasis)

---

33 Figures with identical content are also found in other introductions, including Heitler & London (1927, p. 462) and Heitler (1961, p. 98). Heitler and London's estimation of the binding energy (distance between curve minimum and asymptote) yields 3.14 eV compared with the experimental value of 4.74 eV.

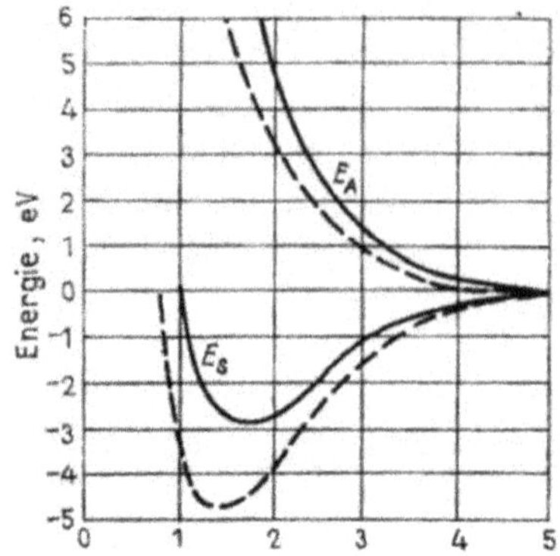

Fig. 21: Energy curves for the hydrogen molecule. The x axis shows Bohr radius units (0,5285 A); the y axis shows the (potential) energy in electron volts. E_S shows the ground state; E_A shows the first excited state. Dashed lines show measured values; solid lines show calculated values. From Ladik (1973, p. 216).

What Heitler refers to as saturation is of course hardly alien to chemistry. The compositional studies of the modern sciences of material substances have long shown that carbon, for example, generally forms quadruple bonds in its compounds (see the exception of the Gomberg radical). However, Heitler does not mean the mere detection of a phenomenon, but its theoretical classification. It is obvious that bonding phenomena cannot be understood without recourse to quantum physics. In conceptual terms, Heitler's saturation is in the vicinity of the octet rule. At least its basic statement is the same, for example for carbon atoms, which in methane form four like covalent bonds with hydrogen atoms and not more. Yet the "chemical" octet concept is alien to Heitler, and accordingly he does not use it in his book. I will conclude this section with a remarkable statement by the theoretical physicist and pioneer of quantum chemistry:

> The philosophically inclined reader may decide himself whether he must regard the wave field of an electron (the ψ-function) as a part of an "objective reality" or "only" a useful product of the human brain that is suitable for predicting the results of the experiments (whereby it becomes clear that these "predictions" invariably coincide with the "objective facts"). The author would prefer to refrain from influencing the reader regarding this point. He merely advises him to first make it clear about what is meant by "objective reality." (Heitler 1961, footnote 12)

8.2.2 Molecular Orbitals

The modern quantum chemical and energetic representation of the formation of molecular orbitals (MO) utilizes a very simple scheme for diatomic structures (see Figure 22). Parallel and antiparallel spin are described with up and down arrows in the orbital box, which may contain at most two electrons. Pauli's principle states that two electrons cannot be equal in all quantum numbers. Therefore, in a

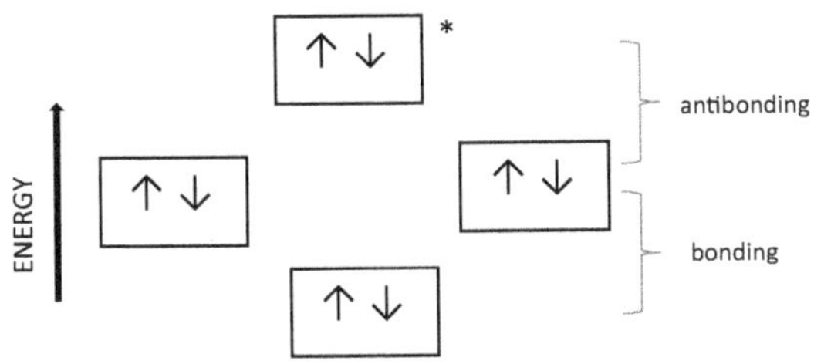

Fig. 22: The molecular orbital model for the (nonexisting) molecule He_2.

fully occupied orbital one arrow must point up and the other down. In a certain sense Lewis's electron-pair theorem is embedded here because there, too, it is always electron pairs that play a key role. There may be exceptions with individual bonding electrons. The classic example of this case is H_2^+, which is also the only two-atomic particle for which there exists an exact solution for the quantum chemical calculations (because it consists of only one electron).

The MO model can be used for example for the discussion of a virtual helium molecule He_2, whereby the total of four electrons, in addition to fully occupying the binding state, would also fully occupy the antibonding state, which explains the nonexistence of this particle (or a substance consisting of it). The information content of such orbital models is significantly greater than in the simple valence structural formula like H – H because what happens in the case of electronic changes can be more precisely described. For example, the excited state is better understood as the shift of an electron from the binding σ-orbital to the antibinding σ*-orbital. The pluralistic coexistence of valence bond and molecular orbital approaches in chemical practice can be explained by the different tasks of the respective models. Valence structural formulas will always be used when the discussion centers around reaction mechanisms, whereas the less vivid MO diagrams aid in understanding spectroscopic phenomena.

8.2.3 Pauling's Ingenious Idea: Electronegativity

The varying capacity or tendency of atoms to make use of electrons in chemical reactions and bonds has been quantified in different variants since the 1930s.[34] The

34 Regarding the almost two-centuries-old history of electronegativity, see the works of Bill Jensen (1969, 2003, 2012). Ruthenberg & Martínez González (2017) examine this topic from the perspective of philosophy of science. Many modern representations follow Pauling and provide an unnecessa-

first such attempt, one which broke new ground, was by Linus Pauling (1901–1994). Pauling, one of the pioneers of quantum chemistry, used the idea of resonance (mentioned above) to represent the "partially ionic" character of a covalent bond. A covalent bond A-B can also be thought of in its resonant (mesomeric) boundary forms $A^+ + B^-$ or $A^- + B^+$. The differing character of the two elements leads to polarity of the bond and Pauling regards this as an additional contribution to the bond ("resonance energy"). It became apparent that the dissociation enthalpy of the substance AB, that is the quantity of energy required to dissociate and separate the constituents, is actually greater than the sum of the average of the dissociation enthalpies of the assumed diatomic constituents A_2 and B_2: $E_D\,(AB) = ½\,ED\,(A_2) + ½\,ED\,(B_2) + \Delta$. This resonance energy Δ is 270 kJ/mol for hydrogen fluoride (HF)and only 4 kJ/mol for hydrogen iodide (HI). With the postulate of proportionality of Δ to the square of the respective difference in electronegativity, these relational values can easily be converted into dimensionless quantities. Pauling defines fluorine, by far the most electronegative element, as $\chi = 4$ so as to obtain positive values for all differences. In this manner the other specified elements receive values of χ such that (H) = 2.1 and χ(I) = 2.5. The differences in electronegativity make it possible to estimate the "ionic character" of compounds. Pauling postulates the following formula for this purpose: Ionic character = $1 - \exp(-¼\,(\chi_A - \chi_B)^2$. Low differences in electronegativity indicate low-polarity bonds (1% ionic character results for $\chi_A - \chi_B = 0.2$), high differences are indicative of a more salt-like character (the greatest possible difference of 3.2 yields an ionic character of 92%). Pauling apparently interprets the term "ionic character" in a realistic sense, although it is difficult to imagine molecules whose reason for bonding is at least partially ionic, that is differently charged constituents that stabilize themselves by means of Coulomb attraction. The charge separation in ionic bonds is generally associated with the image of the ionic crystal and basically conflicts with the molecule concept.

Whereas the Pauling scale of electronegativities was created using the concept of resonance, its empirical basis is thermochemical data, that is dissociation energies measured in samples of material. Electronegativity as Pauling introduces it is not a quantum mechanical concept.[35] This disposition increasingly came to be regarded as a natural law, and, interestingly, all subsequent attempts to create sim-

rily abbreviated definition of electronegativity, as is the case in Riedel (2004, p. 119): "A measure of the capacity of an atom to draw the binding electron pair toward itself in an atomic bond is electronegativity."

35 Accordingly, Preuss (1960, p. 241) discusses Pauling's approach as a "semi-empirical method."

ilar scales for it based on other measured values[36] showed a convergence of the resulting values with those of Pauling. All studies of electronegativity that followed Pauling's introduction refer back to it; more precisely, they confirm Pauling's results.

Reversing in a certain sense the empirical basis just implied, Pauling makes the following statement in *The Nature of the Chemical Bond:*

> The electronegativity scale introduces a certain classification system into inorganic thermochemistry, which at first glance shows hardly any order at all. The electronegativities change in a regular manner from element to element and permit at least an approximate calculation of binding enthalpies. These are [...] by and large identical to the ionic resonance energies that are conditioned by the partial ionic character of the bond between unequal atoms. (Pauling 1962, p. 91)

Here the theoretical construct essentially becomes a natural species, and the estimation of binding enthalpies is granted the same stature as the customary thermochemical measurements.

8.2.4 Hans Primas – an Iconoclast?

Hans Primas has opined on important topics in a series of articles from the borderlands between the theory of chemistry and the philosophy of chemistry. Primas is one of the few quantum chemists with an educational background in chemistry to explicitly discuss the fundamental limits of the application of quantum mechanics to chemistry. Briefly summarized, in his articles he maintains that quantum mechanics brought an ontological sea change into the world that natural science and philosophy must come to terms with. Yet at the same time he feels that the opportunity exists to freely shape natural science within a certain framework.

> The claim that molecules exist in an absolute sense is in any case irreconcilable with the fundamental principles of quantum mechanics. From the standpoint of quantum theory, the object of study in natural science is not undivided reality but the patterns obtained by abstraction. (Primas & Müller-Herold 1990, p. 318)[37]

36 There are experiments to calculate electronegativities from spectrometric data, from dipole moments, and from electron affinities and ionization energies, cf. Ruthenberg & Martínez González (2017).

37 With respect to this publication, an introduction to quantum chemistry, I regard Primas as the senior author and especially as the source of the more philosophical comments, which here are particularly clear and productive. Other pertinent references include Primas (1983, 1985a, 1985b, 1990, 1992). Cf. the article by Amann & Gans (1989), authors close to Primas in terms of content.

The abstractions with which the molecule concept is connected are different from those

> that lead to the concept of material and to chemical thermodynamics. That means quantum chemical molecule theory and chemical thermodynamics are two viewpoints that are mutually *exclusive* yet do not *contradict* each other. (Primas & Müller-Herold 1990, p. 319; original author's emphasis)

Primas thus has a pluralistic perspectivism in mind. In his opinion absolute truth of a scientific statement (except those of quantum mechanics) is unattainable, rather it assumes a sort of empirical adequacy such as that described by Bas van Fraassen. Among other examples from chemistry that are inaccessible to quantum chemistry, Primas mentions

- how many aggregate states of matter there are,
- how a liquid should be characterized theoretically (Primas & Müller-Herold 1990, p. 317), and
- "what the precise connection is between molecules and chemically pure substances (prize question: what exactly does liquid water have to do with the individual molecule H_2O?)" (Primas & Müller-Herold 1990, p. 318).

Primas also touches on the problems of the substance class (Primas & Müller-Herold 1990, p. 311). The approximate quantum chemical calculations for individual molecules such as propanone, butanone, and acetophenone can be readily performed but the classification "ketone" remains inaccessible. Thus it is simply the substance perspective that is inaccessible to the molecular perspective. This is essentially because even as the data for quantum chemical calculations are collected, isolated particles are created and examined. Quantum chemistry stems from nuclear physics:

> The successes of quantum chemistry should not let us forget that quantum chemistry is *a theory of individual molecules* which at best relates to chemically pure substances in an ideal gaseous phase. (Primas & Müller-Herold 1990, p. 312; original author's emphasis)

Thus if it is known which molecule is being studied, then in the best case a more precise quantum chemical mathematical description can be provided. This makes binding energies and certain spectroscopic characteristics accessible. Therefore, this approach is a "quantum mechanical underpinning of chemical experiences."[38]

---

38 I have borrowed this formulation, which describes the situation very well, from Preuss (1960, p. 241). It should be noted that such representations and similar ones occasionally convey (or pos-

How such a molecule of an experimental study can be made accessible is beyond the sphere of influence of quantum mechanics. Material behavior such as the keto-enol tautomerism discussed above remains unaffected. Where individual molecules are examined, macroscopic characteristics remain outside the perspective. Hardy anyone could imagine wanting to calculate the melting point, solubility, or toxicity of a substance based on knowledge of the Hamilton operator of its molecules (if it indeed "consists" of such) together with general natural constants and a few quantum numbers.

Hans Primas is decidedly pessimistic regarding a theory of substance chemistry:

> Doing without a doctrine of substance anchored in chemical phenomenology led to a concentration on the study of the so-called elementary building blocks and has released unimagined energies and achieved spectacular results. However, one reason the successes of the molecular view appear so spectacular is because the alternatives have been crippled. (Primas 1985b, p. 163)

The description of the predominance of molecular theories in chemistry (and now in biology as well) certainly continues to be correct. Just think of the usual publication strategy: Without a section with quantum chemical calculations it has become nearly impossible to get an article motivated by a synthetic line of inquiry through the peer review system. In this manner an entire industry of activities in chemical theory has arisen, and one has the impression that the calculations have become more important than laboratory results (if indeed there are any worthy of the name). And the powerful computer workstations are making it increasingly appealing (and inexpensive for the institutions) to practice chemistry solely "on screen" as it were.

Primas is completely correct in how he describes the predominance of molecular quantum theory models that have arisen from the "triumphant success" of atomism and electronism, at least as far as conventional papers for a scientific audience are concerned. Yet seen in the light of day substance chemistry is not yet lost; indeed quite the opposite is true: The actual chemical work in laboratories, workshops, and companies does not generally take place at the submicroscopic level but uses perceptible, weighable substance specimens. Even the most beautiful brightly colored graphics in our publications and the most abstract calculations

---

sibly are intended to convey) the impression that the words "quantum mechanical" impart the content a particularly high degree of scientific reliability and significance. At the same time the relationship between empirical background and theory is often nearly turned upside down when it is repeatedly emphasized that quantum mechanics is among the physical theories best confirmed by experimental findings, whereas its origin from spectroscopic experience is practically ignored.

have their correlates in what are usually rather mundane, occasionally foul-smelling objects which may require complicated cleaning operations before they can be made into something one would want to write about. And as we have read above: For whatever reasons we may practice chemistry, the production aspect can never be fully ignored. Chemical theory, even if it includes nuclear physics and quantum mechanics, is not superior to empirical work but subordinate to it. And as a rule this empirical work is work on substances not molecules, much less individual molecules. Although a strict theory of substance, say in the form of a mathematical algorithm, does not in fact exist, there are 1) reasons for this (degree of complexity, inability to render abstract) and 2) sufficient indications, in the writings of Wilhelm Ostwald and František Wald (both of whom Primas mentions), of which aspects could be the central ones here (see the respective chapters). Logically, Primas calls for theoretical pluralism or, respectively, complementarity for chemistry (and for other areas as well):

> The molecular description of matter is compatible with quantum mechanics and, as far as we know, empirically correct, but it captures only a fraction of the nature of matter. We need more and more fundamentally different descriptions and complementary standpoints in order to make all of reality discernible [...]. A current and urgent task of theoretical chemistry is therefore to develop the various complementary viewpoints, clarify the mutual dependencies of complementary descriptions, and invent new descriptions. (Primas & Müller-Herold 1990, p. 319)

8.3 Further Limits of Quantum Chemistry

The more general limitations (with respect to the chemistry of substances) of energetic and quantum physical approaches in chemistry may be said to include the fact that they are largely inapplicable to kinetic aspects of chemistry (speed and mechanisms of reaction). As in quantum mechanics, where the only existing analytical solution to the Schrödinger equation is for the chemically irrelevant particle H_2^+, in kinetics as well absolute calculations are only possible in extremely artificial cases that are far removed from any chemical relevance. As I have shown in the respective chapter, the temporality and potentiality of chemical processes are fundamentally different conceptional perspectives. The Schrödinger equation is a genuinely energetic concept that includes kinetic energy and thus time as a value. Yet this kinetic energy refers to the electrons within the molecule and not to the course of time in a chemical reaction that this molecule may pass through. An indication of the fundamental separation of chemical kinetics from a quantitative quantum chemical interpretation is the fact that the first major textbook on the "theory of the transitional state" or even the "theory of absolute reaction speeds" contains an extensive introductory chapter on quantum mechanics. Its application thereafter is primarily limited to the assumption that the atomic nuclei move slowly relative to the electrons involved, meaning that the particles are in the ground state. Such a state is referred to in quantum mechanics as "adiabatic." For nearly a century the model of potential surfaces ("Born-Oppenheimer hypersurfaces") has been a standard tool for what amount to classical explanations in chemical kinetics, and it permits a number of enlightening estimates.[39] However, the crucial set pieces of the theory of the transitional state come from static thermodynamics, not from quantum mechanics.

Together with members of her research team, the Argentinian philosopher of science Olimpia Lombardi has been examining specific problems of the use of quantum mechanics in chemistry for some time. A few of her results bear mention here. In a paper about the relationship between molecular structure and quantum mechanics, the authors maintain that the spatial structure of molecules cannot be calculated absolutely by means of quantum mechanics (Martínez González et al. 2019). As their main argument, they assert that distinguishable elementary particles do not exist for quantum mechanics and therefore neither does the spatial arrangement of individual atomic nuclei relative to one another. They also emphasize that the additions of "classical" approximation assumptions such as the Born – Oppenheimer approximation do not always harmonize with strict formal quantum mechanics, at least when one wishes to speak of a "physical underpinning" of chemistry, in other words, its theoretical reducibility.[40] Another article by this group (Arriaga et al. 2019) examines the theoretical approach of the Canadian chemistry theorist Richard Bader (1931–2012), the *quantum theory of atoms in mol-*

---

39 The textbook in question by the Princeton chemists Glasstone, Laidler & Eyring (1941) contains the original graphic of the potential surface representation on page 96. Primas & Müller-Herold (1990) also reproduce this figure (1990, p. 202) and briefly discuss "adiabatic reactions" although this is not a genuinely quantum chemical topic. In their summary they state (1990, p. 205): "Often even a contemplation of the plausibility of the structure of the transitional complexes allows interesting qualitative statements to be made about the possible course of chemical reactions." Hinne Hettema (2012) discusses the case study of the "absolute reaction rate theory" from the standpoint of the philosophy of chemistry. The author is receptive to reductionism in the sense of a path to a unified natural science and regards the theory of the transitional state as a typical example of this attitude.

40 Other authors also come to a similar conclusion. Paul Bogaard for example writes in a brief but very informative article about the influence of Gilbert Lewis on Linus Pauling and Charles Coulson about the latter's opinion (Bogaard 2003, p. 308): "Molecular structure – and the search for directionality of bonding that could produce it – is motivated neither by the mathematical formalisms of quantum theory nor by the physics which underlies quantum mechanics, but by experience with chemical substances."

ecules (QTAIM) as the author himself described it. In this recent model for describing chemical bonds, greater emphasis is placed on relationships of nuclear physics than was previously the case with the valence bonding approaches and the molecular orbital model:

> The quantum theory of atoms in molecules (QTAIM) uses physics to define an atom and its contribution to observable properties[41] in a given system. It does so using the electron density and its flow in a magnetic field, the current density. (Bader 2013, p. 253)

The advocates of this model hide neither their extreme reductionist attitude nor their equally extreme physicalistic-naturalistic attitude:

> A scientific discipline begins with the empirical classification of observations. It becomes exact in the sense that predictions are possible as soon as the classification reflects the physics underlying an observation. (Bader et al. 1994, p. 647)

The actual classification in the science of substances initially proceeds from the relational reaction behavior of samples of material and not via a theoretical comparison with current theories of physics let alone the respective calculations. Nobody will actually calculate what a "ketone" is and whether or not a substance specimen to be examined belongs to this substance class. Naturally this does not mean that such calculations would be completely useless. It merely strikes me as strange to immediately brush aside successful strategies for characterizing material when there may be an opportunity to "reflect" physics. Here, too, it must remain clear that the theory, whatever theory that may be, is secondary to the empirical work (for this reason I have no quarrel with the first sentence of the last quotation). If one desires to draw ontological and metaphysical conclusions – such as the conclusion that priority should be given to a certain theory and its simplifications as well, thus immunizing it in a certain sense against empirical findings – then, at least from a philosophical perspective, one becomes forced to provide an explanation.

The group around Olimpia Lombardi argues against the QTAIM in a manner similar to the previous example of the molecular structure. Again it is correctly pointed out that the individualization of particles conflicts with all conventional interpretations of quantum mechanics.[42] Another article by the Argentinian

41 By observable they mean somehow physically tangible and therefore measurable values.
42 They note that the Bohm variant of quantum mechanics has better chances of being compatible with the QTAIM but they do not expand on this idea (Arriaga et al. 2019, p. 135).

group examines optical isomerism (Fortin et al. 2016). This expression is used to describe mirror image isomerism, as to, for example, lactic acid:

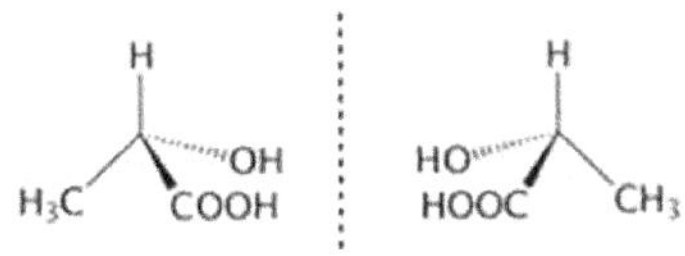

Mirror image isomers (enantiomers) turn the plane of oscillation of the polarized light in opposite directions but otherwise show practically identical chemical behavior. An important practical aspect also suggests that the two forms are not identical as substances. If one mixes samples of both, the mixture will behave exactly like mixtures of normal positional isomers: For example, melting points are lowered. Ostwald's classic regulation for the identification of substances applies here, too, even if his "substance law" (two samples are of the same species when they share a few common characteristics) is not easily applicable. Discussing this phenomenon from a quantum chemical standpoint involves a specific twist: the Hamilton operator is the same for both antipodes. The authors note this as they do that fact that the Born – Oppenheimer approximation cannot have any influence on the interpretation in this case. Friedrich Hund (1927) examined the so-called "paradox of the optical isomers." This paradox describes the fact that the existence of the two stereoisomers cannot be differentiated by quantum mechanics (nor by thermodynamics). Synthesizing a potentially optically active material without specific stereo-selective measures yields a "racemate," that is a mixture of both antipodes. As the two are energetically equivalent there must be other reasons for this if it is possible to produce or separate pure isomers (stereo-selectively), and these reasons can be classified as belonging to the realm of chemical kinetics.[43] In a certain sense it is a specious problem, one that Hund, whom Fortin et al. discuss, seeks to solve with the following approach:

> We must show that quantum mechanics finds the life span of the left-handed or right-handed states to be exactly as great as the findings of the experimental study of mirror image isomers. (Hund 1927, p. 811)

Because the stereoisomers are not generally in equilibrium with each other, the study of "beat periods" is a futile pursuit. Thermodynamically, the two antipodes have absolutely identical stability and even with the decoherence concept, that is essentially the approximation of the real chemical environment, one major

43 I see this as further support for my thesis that energetics and kinetics are fundamentally different.

area of problems in quantum chemistry will certainly continue to be in integrating in a practicable manner the classic characteristics and values (possibly observable) from the macroscopic substance perspective, such as mass, temperature, chirality, molecular structure, chemical potential, etc.[44]

Hans Primas, in the *epilogue* of the introductory textbook of quantum chemistry repeatedly mentioned here, makes the following statement:

> The most important thing is the realization *that there are no specifically chemical forces.* The essential cause of the chemical bond phenomenon is the electrostatic Coulomb attraction between the negatively charged electrons and the positively charged nuclei, in combination with the laws of a new mechanics. (Primas & Müller-Herold 1990, p. 301; original author's emphasis)

Accordingly, the postulate of a "chemical" form of motion propagated by Friedrich Engels and other dialectical materialists must at the very least undergo a thorough "de-ontologizing" revision. Chemistry will presumably continue to get along well with the fact that it will generally have to do with middle-sized objects and longish processes and not with the ultimate building blocks of the universe or primordial matter.[45]

8.4 Conclusions

From both a historical and a systematic perspective, quantum chemistry can be reduced to the study of isolated atoms and molecules. As a result, quantum chemical calculations in the key field of chemistry, the behavior of substances, are only applicable to a limited extent.[46] All chemically interesting quantum chemical calculation methods are approximations. A genuine ab initio method, that is calculations relying only on natural constants and mathematical algorithms, does not exist.

---

44 The examples listed come from Primas & Müller-Herold (1990, pp. 125 and 197).

45 Here I have borrowed part of the title of the book *Macroscopic Metaphysics – Middle-Sized Objects and Longish Processes* by Paul Needham, which is in fact a work in the philosophy of chemistry (Needham 2017). Cf. Ruthenberg (2016a, 2020c).

46 In the textbook *Philosophie und Naturwissenschaften* by a group of former East German authors, we read: "The achievement of quantum theory as applied to chemical problems consists not only in having provided a more adequate theory of the chemical bond but also in the realization that the laws of the macro level cannot be directly applied to the micro level." (Author collective 1988, p. 189). I agree with this, although I note that the reverse statement is also true: it is equally impossible to apply the laws of the micro level to the macro level.

The empirical findings around the "discovery" of electrons accelerated and intensified the acceptance of atomism, ironically with the proof of the divisibility (ability to form ions) of atoms and accompanied by theoretical uncertainty about what electrons actually are (wave-particle dualism). The particular significance of the electron as an epistemic object, especially in the theory of the chemical bond, was recognized, and occasionally exaggerated, prior to the development of quantum chemistry.

Quantum physics has developed from a theoretical superstructure with epistemic underpinnings into a metaphysical foundation of work in the natural sciences, and this foundation is often no longer critically examined. As this has occurred, important chemical concepts such as the octet with the octet rule and the oxidation number have seemingly fallen by the theoretical wayside. Regardless of this development, various bond models such as the valence bond method and the molecular orbital approach continue to be used in parallel. Even in this field of endeavor in the science of substances, a pluralism based on pragmatism predominates.

In general, at least to acquire an initial impression of a chemical phenomenon, the following empiricist principle should apply: Conclusions about substances should be drawn from experiments with substances; conclusions about atoms and molecules should be drawn from experiments with isolated atoms and molecules; and conclusions about electrons should be drawn from experiments with electrons. Many mereological projections are possible, but switching between categories requires an explanation. The closer we want to get to the non-observables, the more we move away from the metachemical foundations and approach metaphysics.

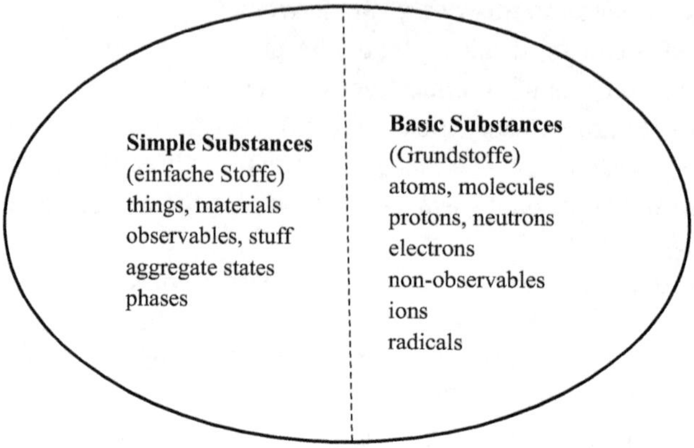

Fig. 23: Schematic diagram of Friedrich Paneth's dualistic element concept, consisting of the real-world empirical *simple substances* and the theoretical "transcendental" *basic substances* with their nuclear charge (number of protons in the nucleus), understood as an essence.

9 Friedrich Paneth – a Philosophizing Chemist

There are not many chemists who reflect on their own science from a philosophical perspective. Born in Austria, Friedrich Adolf Paneth (1887–1958) belongs to this particularly small group. He has left his mark on both the science of chemistry and the philosophy of chemistry. I will examine this mark in four sections, which I order roughly according to the historical sequence of their origin.

9.1 A Conceptual Crisis in Chemistry, Set in Motion by Physics

In 1916 Friedrich Paneth from the Institute of Radium Research in Vienna published what can be called a methodological paper the essay *Über den Element- und Atombegriff in Chemie und Radiologie* [On the Concepts of Element and Atom in Chemistry and Radiology] (Paneth 1916). This essay examines crucial concepts that, first, have a complex prior historical development, second, exhibit different standpoints in allied disciplines, and, third, whose handling has apparently been caught in a crisis as a result of empirical experience.[1]

In this essay Paneth explores the topic of the atomist definition of the chemical element as initially formulated in modern science by Robert Boyle (1627–1691) and then championed by John Dalton (1766–1844). This classic definition basically states that there are as many chemical elements as there are types of atoms. And all of these atoms, Dalton stipulates, are to have the same weight. Against the background of contemporary empirical results, namely of the decay of elements (Rutherford & Soddy 1902), of atomic structure (Rutherford 1911, Bohr 1913), and of isotopes (Soddy 1913) of many elements, the question arises for Paneth as to "whether the law of the equal number of elements and types of atoms retains its validity today" (Paneth 1916, p. 171). In his historical outline, he presents Antoine Laurent Lavoisier's (1743–1794) rejection of the speculative notion held by Empe-

1 The term "radiology" in this context is not understood to mean the medical application of radiation in the diagnosis and treatment of disease, but the science of radioactivity. After receiving his doctorate in organic chemistry in Vienna in 1910, Paneth joined the Institute of Radium Research in 1912 and in so doing fundamentally changed the direction of his work (although the synthetic aspect of his research remained, see below). He became a pioneer of radiochemistry, at the time a new field, and collaborated closely with György József von Hevesy who received the Nobel Prize in 1943 (in part for initially collaborative work). Paneth organized his essay as follows: 1. Necessity of a revision of the definition, 2. Historical retrospective, 3. The most important discoveries of recent years related to the element and atom concept, 4. The standpoint of chemistry, 5. The standpoint of radiochemistry, 6. Final remarks, 7. Conclusion. For a short biographical account see Ruthenberg (1997).

https://doi.org/10.1515/9783111401041-012

docles and Latin alchemy that the number of elements is limited as well as Lavoisier's operational definition of element that was influenced by Boyle. In this definition, elements are substances that cannot be further separated into simpler substances by any known manipulation, i.e., one that may be regarded as "chemical." Lavoisier (1803, pp. 12–13) states: "They appear to our eyes as simple bodies and we may not regard them as composites until the moment that experience and observations have given us evidence thereof." Paneth (1916, p. 174) sees the drawback of this operationalist view in "provisionally sacrificing a basic principle of explaining nature, [namely] reducing qualities to quantities [...]." In contrast to Lavoisier, Boyle had exactly this in mind earlier: With the aid of the atoms distinguishable in form, size, and motion envisioned by Leucippus and Democritus, Boyle aimed to reduce the substance qualities to quantitative differences. But only with the assumption of qualitative differences of atoms as Dalton introduced around 1808 did it become possible, as Paneth states (1916, p. 175), for "Boyles' definition of chemical elements to fully develop its fruitfulness." And in the same place he continues: "The goal of universal philosophical explanation of the world was thus initially cast aside, yet this laid the foundation for the grand development of chemistry."

The observation of the possible decay of elements in the early phase of radiochemistry was explained by the hypothesis that the atoms of a radioactive element are transformed into other atoms, that is atoms of another element.[2] The discovery of isotopes,[3] in modern terms the occurrence of many chemical elements having different sorts of atomic nuclei (nuclides) which nonetheless maintain their chemical character, is among the important discoveries which according to Paneth are related to or have influenced the concept of element and atom. The third relevant scientific aspect Paneth names is what is known as Moseley's law (1913). When he examined the relationship between the frequency of the radiation emitted by the anode material of elementary metal and its position in the periodic table in an x-ray experiment, Henry Gwyn Jeffreys Moseley (1887–1915) found that assuming atomic weight as a criterion for ordering the elements is incorrect. It was the atomic number, more precisely its square, and not the atomic weight that proved to be proportional to the x-ray frequency ($Z^2 \sim n$). Paneth emphasizes as a unique feature of Moseley's work the fact that it

2 These were analytical chemical processes for identifying substances and elements from which the atomic realm was inferred without direct empirical findings. The phrase element decay, which we may classify as belonging to the substance level of knowledge, is thus correctly chosen.
3 Isotope = "in the same place," refers to the position in the periodic table of the elements; the term was introduced in 1913 by Frederick Soddy (1877–1956).

> demonstrated the possibility of a strict systematic classification of the elements based on their high-frequency spectra and thus simultaneously cast a bright light on the significance of the old chemical types that are greatly limited in number, in contrast to the swarm of newly discovered radio-elements that expands daily. The old concept of the element received its strongest support from them. (Paneth 1916, p. 180)

In contrast to the linear structural principle that results from Moseley's law, the periodic notion introduced by chemists has the enormous advantage of identifying both horizontal relationships (periods) as well as vertical relationships (groups). Paneth therefore concludes that the periodic system is superior to the linear one in its predictive power. Paneth views the new scientific findings in the following section from a chemical and radiochemical perspective. He finds that the fact that atoms can decay, that is the phenomenon of radioactivity as such, does not affect the validity and applicability of the atom concept, especially in the chemical range of experience. As a matter of principle, we can no longer proceed from the immutability of the elements in light of the new radiochemical findings. Yet how should we handle isotopes? The existence of isotopes clashes with Ostwald's substance law, which Paneth expressly refers to as a rule.[4] This states that substance samples that exhibit a few matching characteristics are identical in all their characteristics. Paneth maintains that this law-like relationship has been rendered nugatory by the discovery of isotopes and gives as a reason the possible fluctuations in atomic weight in the case of certain elements. In the discussion of this situation he arrives at the crux of his study: How far do we want to go with the element definition?

> From a strictly logical standpoint, both ways are possible here: postulating that all characteristics are identical or allowing a range of fluctuation for a few. However, closer examination will show us that the first alternative, which at first could seem to be more reasonable, leads to complete debasement of the element concept for chemistry. (Paneth 1916, p. 182)

In order to avoid such a debasement of the element concept, Paneth opts for the second way which makes do with identifying the chemical characteristics and formulates the following definitions:

> An element is a substance that cannot be further separated into simpler substances by any chemical process. Substances that comply with this definition are regarded as the same element if they cannot be separated by any chemical process once mixed with each other. Atoms

4 Paneth refers to Ostwald's *Grundriss der allgemeinen Chemie* of 1899. Ostwald discusses this in detail in his *Prinzipien der Chemie* (see Chapter 4).

> are the building blocks of matter to which chemical separation can advance but which themselves remain intact in all chemical reactions. (Paneth 1916, p. 198)

The first part of this conclusion corresponds to the traditional definitions by Robert Boyle and Antoine Lavoisier. It would certainly be plausible and operationally feasible to consider the fact that there is a consensus about the "chemical" processes.[5] However, in the second part of this conclusion, which relates to atoms, Paneth is leaning very far out the window: To advance "chemical separation" down to atoms is not feasible in the strict sense, rather it is more of a thought experiment. Although specialist jargon may suggest otherwise, identifiable individual atoms are worked with in chemistry only in theory, never in reality. However, one need not necessarily interpret Paneth's statement in the operationalist sense that such a separation is possible, but better in the conceptual sense that such an ongoing separation of a substance sample by whatever means would end at atoms. In any case it is an abrupt category shift when Paneth on the one hand refers to the *substance* law yet on the other hand treats atoms as a possible substrate and even cites the discovery of isotopes as a reason for the invalidity of Ostwald's description: Even in the early phase of radiochemistry it had become clear that radioactivity is not a substance characteristic (although many specialists even refer to the field as "nuclear chemistry"). Although the substance law most certainly has its restrictions and the necessary conventions must be carefully observed, it remains a useful metachemical premise for chemical work, whether practical or theoretical.

In his 1916 article, Paneth describes the suggestion he presents as an "empirical definition for the chemist's practical purposes" (Paneth 1916, p. 193) and in the same passage admits that not all features or characteristics (such as characteristic spectra and fluctuating atomic weights) can be derived from this definition. He states:

> One could only comply with this requirement – derivation of the consecutive features from the constitutive one that has been incorporated into the definition – by basing the definition itself on a theory of atomic structure. From the standpoint of the *Rutherford-Bohr* model of the atom, such a theoretical definition would have to look like this: *An element is a substance all of whose atoms have the same nuclear charge.* (Paneth 1916, p. 194; original author's emphasis)

5 What counts as a "chemical" process is just as dependent on context as the meaning of the expression "chemical characteristics." Paul van der Vet noted this aspect in his meticulous study; cf. van der Vet (1987, pp. 81–82). The latter study is the only one to date that also discusses in some detail Paneth's crucial 1916 essay examined here.

This is the point at which Paneth first refers to a (in his words) "theoretical" element definition and contrasts this to the "practical" version.[6] Both have advantages and disadvantages but are not mutually exclusive; thus they can apply simultaneously, depending on the chosen perspective. The practical definition is thus the "chemical" one, the theoretical definition the "physical" one. The classic dichotomy ontological/epistemological appears applicable with a clear result. Just as a classification of the theoretical version appears to be the more fundamental to the typical essentialist, the practical one appears to the descriptivist as the fuller and more adequate one because it is substance-related. For Paneth, the essential purpose of his 1916 essay lies in making the case for a practically manageable definition (especially in chemistry) and to prevent each newly discovered isotope from being designated an element in the future. He succeeded in the latter although this initially required him to endure polemic onslaughts from some of his colleagues. For example Rudolf Wegscheider of the University of Vienna refers to Paneth's expressly preferred practical and non-atomic part of the element definition and states:

> The definition of elements given by Paneth does not meet the needs of chemists. The chemist primarily has an interest in [seeing] that all characteristics relating to the inner nature of a body are clearly defined by specifying its composition, and must therefore base the concept of substance on the complete identity of all characteristics: He must therefore postulate as many substances as are necessary for this purpose, and consequently he cannot regard isotopes as a substance. (Wegscheider 1918, p. 749)

Wegscheider obviously feels that the atomic weight (atomic mass) also belongs to the chemical characteristics or, better, to the substance characteristics. However, a consequence of Wegscheider's rigorous approach would be that the conceptual differences between physical matter and chemical substance would be fundamentally obscured, exactly what Paneth seeks to avoid.[7] Consequently, Paneth excludes the fluctuations in atomic mass due to isotopes as irrelevant to the definition of an element.

In his review of Kasimir Fajans' 1919 book *Radioaktivität – und die neueste Entwicklung der Lehre von den chemischen Elementen* [Radioactivity and the Latest Development of the Theory of Chemical Elements], Paneth deals nearly exclusively

6 Interestingly, Sarah Hijmans (2020), because she does not refer to Paneth's earlier writings, arrives at a similar conclusion when she "suggests" the interpretation of this dualism as the empirical and theoretical view of elements.

7 The main opponents of Paneth's proposal – proceeding from different points of criticism – were Heinrich Remy (1890–1974), Rudolf Wegscheider (1859–1935), and Kasimir Fajans (1887–1975). Cf. the study by Paul van der Vet (1987).

with the criticism of his own position. He writes about the author: "He defines a chemical element as a substance that cannot be separated into simpler components by any means and has not been recognized as a mixture of other substances." Then he refers to Rutherford's phenomenon of breaking down nitrogen and concludes that according to Fajans this would mean that nitrogen is no longer an element. Paneth continues: "[...] yet this conclusion is rejected by the author himself, leaving no other choice but to give up the definition" (Paneth 1920, p. 94). He then recapitulates the two versions of the element definition (the *practical* and *theoretical* ones) and again criticizes the "isotopic" view of elements, that is to consider every nuclide a separate element. Paneth's approach is plausible: Although modern chemistry refers often and extensively to protons, (characterization of elements, periodic table), electrons (theory of bonding, reactivity, and affinity), and later neutrons as well (stoichiometry, isotopes), they are not directly accessible with the analytical and synthetic methods and tools of the substance-oriented chemical sciences; they thus remain non-observables (or theoretical entities). As has been implied, the scientific community, in the form of the newly established *International Union of Pure and Applied Chemistry* (IUPAC) founded after the war in Rome in 1920, accepted Paneth's suggestion at least in part. As a result of how the war ended, Paneth himself, like all other German and Austrian researchers, was excluded from participation in the relevant bodies.[8] One example of the acceptance of Paneth's concept is the following conclusion of Cambridge scientist Francis William Aston (1877–1945),[9] who compares the isotopic and chemical versions of the definition. Regarding the latter, he states:

> The second [alternative] was to fix the word [element] precisely, now and for the future, as meaning a substance with definite chemical and spectroscopic properties which may or may not be a mixture of isotopes – in other words, to associate it exclusively with the conception of atomic number. The second alternative has now been generally adopted. (Aston 1942, p. 217)

Nonetheless, chemical properties are still referred to here. The chemical and operational (or "practical") definition that goes back to Boyle and Lavoisier and is still used in practical chemistry (in stark contrast to the "theoretical" definition) has since been increasingly neglected or simply ignored. Even in his later publications on the problem of chemical elements, Paneth makes every effort to draw attention to what he feels is this fundamental dualism. In the *Lehrbuch der Radio-*

8 Cf. van der Vet (1987, pp. 106–109).

9 Aston played a major role in developing mass spectrometry and received the 1922 Nobel Prize in chemistry. He was a member of the first commission for chemical elements of the IUPAC.

aktivität, which he coauthored with Georg von Hevesy,[10] Paneth formulates his "practical" alternative as follows:

> *A chemical element is a substance that cannot be further separated into simpler substances by any chemical process.* (Hevesy & Paneth 1923, p. 123; original authors' emphasis)

In the same passage he emphasizes the significance of maintaining Boyles' notion of the chemical indestructibility of the elements. As an example of the non-chemical breakdown of elements, he mentions the splitting of aluminum atoms by bombardment with α-rays. In the corresponding footnote he continues in a programmatic manner:

> With all chemical reactions we can be certain that they neither "split" atoms nor "unmix" isotopes; the compoundedness – present in a double sense – of the elements can for this very reason be ignored in chemistry. (Hevesy & Paneth 1923, p. 124)

The "practical" element definition has the enormous advantage that according to Ostwald's substance law, and all of Paneth's criticism notwithstanding, no specific measuring methods are required to design a workable process of identification. The "theoretical" definition is explicitly atomist and because it is "farther from the chemical sense of the element concept" it is accessible experimentally only indirectly via highly specialized instrumental physical processes (and only with suitable integration of synthetic steps, see above).

9.2 Cold Fusion and an Undetected Blank Value

Even in the publication of his lectures as a George Fisher Baker Non-Resident Lecturer at Cornell University in 1926–1927, Friedrich Paneth, at the time a professor at Berlin University, took the opportunity to present his dualistic view of the chemical elements and he repeatedly mentions the "practical" version first (Paneth 1928, pp. 139–140). However, in this section I would like to examine another episode that took place at the same time and that has not yet been discussed in the literature on the philosophy of chemistry.

An article published in quick succession in *Chemische Berichte* and then in *Naturwissenschaften* under the simple title *Über die Verwandlung von Wasserstoff in*

10 Von Hevesy did not show any great interest in philosophical issues and appeared only occasionally as a coauthor in the early phase of the disputes about the element concept (cf. van der Vet 1979).

Helium [On the Transformation of Hydrogen into Helium] appears similar at first glance to any other report from synthetic chemistry, and the article does indeed have a synthetic intent. The authors describe their task as experimental verification of the possibility of forming at least small quantities of helium from normal hydrogen in or on palladium as a catalyst. One underlying idea for this is that the necessary reaction can be regarded as energy-producing (exothermic) and that the goal must be to trigger or activate the reaction, understood as an analog of a chemical process, in an appropriate manner. Palladium, a metallic element of the platinum group first described in 1803 by William Hyde Wollaston, can absorb enormous amounts of hydrogen (Graham 1869). Knowledge of this behavior led to the notion that the hydrogen particles must come very close together in the solid state. In which form the hydrogen is present in the palladium, whether grouped as molecules, atoms, or otherwise, is still unknown at the time in question. As they begin their work, the authors are not yet certain as to where the acceptable detection limit for measuring helium must lie: "Yet today after having studied this problem for several years we believe that we have successfully solved it, and we would like to report briefly on this in the following section." (Paneth & Peters 1926, p. 957)

Preparation of the sample gas mixture for the emission spectrometry used for the measurement proceeds as follows: easily condensed gases are removed over active charcoal cooled with flowing air; hydrogen is burned with excess oxygen on a platinum or palladium contact; the remaining oxygen is adsorbed onto cooled active charcoal; the remaining gas is introduced into a glass capillary tubing with an internal diameter of 0.1 mm (this tubing, wrapped with an excitation spool, is used directly for measuring); the setup exploits the circumstance that in the case of longer non-electrode (inductive) discharge, the undesirable accompanying gases (primarily H_2) are broken down more quickly than the helium and neon that, according to the authors, remains. The presence or absence of neon in this experiment has crucial methodological significance:

> For the occurrence of neon is a highly valuable indication that atmospheric air has penetrated; the danger that this occurs as a result of some deficient stopcock or a weak spot in the glass is so great that with a minimal trace of helium one can only negate its origin from air with certainty if the absorption agent is adjusted so that in this case neon would also have to be present in the spectral tube along with the helium. (Paneth & Peters 1926, p. 957)

The authors estimate the detection limit using a thorium preparation which they got from Otto Hahn and Lise Meitner of the Kaiser Wilhelm Institute of Chemistry. It is specified at a minimum of 10^{-8} cm^3 (10^{-12} g) of helium. The green helium line with λ = 5016 Ångström is used for this estimation. After excluding the possibility of helium formation due to electrical discharge, Paneth and Peters describe the use of different qualities of palladium for their experiments, initially common wire, then palladium black, palladium sponge, and palladium asbestos.[11] In this manner, they succeed in “manufacturing preparations that after 12 hours when heated simultaneously with hydrogen gave off enough helium to display 4–5 lines of the helium spectrum without a neon line” (Paneth & Peters 1926, p. 959). Whereas the experiments with palladium wire do not proceed very convincingly, when using specimens with greater internal surface area they have the impression “that the proportionality between effect and duration of experiment that was missing earlier was now present at least roughly.” With palladium asbestos, which works very well, they succeed in detecting 10^{-8} to 10^{-7} cm^3 of helium about twenty times daily before the specimen becomes ineffective.

> Naturally, we have repeatedly posed the question of whether the occurrence of helium in our apparatus can be explained by helium already present. (Paneth & Peters 1926, p. 960)

The authors again emphasize the important conclusion that positive evidence of neon would allow: It is essentially a marker for air or contamination. In the same passage they discuss the selective permeability of glass for helium as opposed to neon. A second possibility for “fractionated separation” of helium and neon results from the disposition of glass to absorb helium in certain quantities but not neon. They also report on the negative course of experiments on the absorbability of palladium and asbestos with respect to helium and neon, and report on experiments on the outgassing of helium as a result of postulated radioactive decay (with emission of α radiation) on old sheets of palladium dissolved in nitric acid.

Thus with all due care and caution, Fritz Paneth and Kurt Peters come

> to the conclusion that the helium occurring in our apparatus owes its origin to the influence of the palladium on hydrogen. Specifically, the failure of helium formation when the palladium is charged with oxygen and its recurrence when charged with hydrogen seems to allow us no other interpretation. For even if one […] would want to assume that the palladium contains helium, one would have to add the further hypothesis that this helium can only be released from it by hydrogen. (Paneth & Peters 1926, p. 961)

---

11 What is known as “black” (German: “Mohr”) is a fine metallic powder obtained by subjecting usually complex salts to extreme heat and then reducing them in a stream of hydrogen. A “metal sponge” is created when a metal salt is heated on a substrate material such as paper or ceramics. Here, too, finely distributed metal forms with an extremely elevated surface. Metal asbestos is metal precipitated in a similar manner on an asbestos substrate. Normal palladium can take up a quantity of hydrogen 900 times its own volume, palladium black 1200 times its own volume.

This interpretation (as we know today) is wrong. As proves to the case just a short time later, nearly all component data for a correct conclusion are contained in the original article; only the causal allocation is still inappropriate, that is to say, the interpretation of the observations and measurements is inadequate.

Let us begin by having a look at the conceivable reaction mechanism. Whereas all descriptions or mentions of so-called "cold fusion" from 1989 on proceed from deuterium as the initial material, the Berlin group in 1926 is apparently considering the possibility of converting "normal" hydrogen (protium, ^{1}H) into helium. In modern notation and using the most stable respective isotope, one could attempt to describe the corresponding process as follows:

$$^{1}_{1}H + ^{1}_{1}H + ^{1}_{1}H + ^{1}_{1}H \rightarrow ^{4}_{2}He \tag{1}$$

Assuming that the hydrogen in palladium is distributed in the form of individual atoms, that is homolytically dissociated as it is introduced ($H_2 \rightarrow 2\,H$), then four initial particles must react with one another in this process, which is practically impossible simultaneously and would also be highly improbable as the result of a reaction or impact. Additionally, two of the protons from the hydrogen atoms would have to be transformed into neutrons. From the perspective of our current understanding, this process would be nothing less than mysterious. If one were to consider molecular hydrogen (H_2) as an initial material, then the reaction would be significantly clearer (yet describable neither in conventional chemical notation nor in nuclear notation and thus also incorrect, like the case above):

$$H_2 + H_2 \rightarrow He \tag{2}$$

In this case, we would have to explain how the fusion of the two two-atom molecules would occur (and, as in the first case, how the proton transformation would occur). With respect to the mass, this process is analogous to the following reaction:

$$^{2}_{1}H + ^{2}_{1}H \rightarrow ^{4}_{2}He \tag{3}$$

Yet in the year 1926 Paneth and Peters were unable to pose such a question because it was only in 1932 that James Chadwick (1891–1974) first reported on the existence of neutrons (Chadwick 1932). The fact that the electron counts do not balance in reactions (1) and (2), if one understands them in an atomic sense beyond the purely nuclear reaction, is less critical; electrons can be regarded as uninvolved in terms of nuclear physics, and in transmutations their counts would necessarily have to be nonzero. The fact that the electron count is neglected should have appeared incomprehensible to the chemists working here – that is if they had stopped to think about it. They appear to proceed from an atomic mechanism.

For example, they refer to the "unification of hydrogen atoms" (Paneth & Peters 1926, p. 961).

In any case, it becomes clear from this brief description and comparison with equation (3) why modern thinking on "cold" nuclear fusion (since 1989) has proceeded from deuterium:[12] A conventional two-particle reaction instead of the four-body problem and the simple addition of the nuclear particles instead of a mysterious transformation of nucleons appear significantly more promising. Because the quantities of helium detected by the Berlin chemists are extremely small, it cannot be excluded that their occurrence might be the result of the reaction of small quantities of deuterium naturally present in the hydrogen used. Paneth and Peters naturally do not discuss this possibility because they do not yet know anything about deuterium: Its lines in the spectrum of hydrogen were only described in 1931 by the American chemist Harold Urey (1893–1981), for which he was awarded the Nobel Prize in chemistry in 1934.[13] In any case, the Berlin researchers do not speculate at all about possible elementary reaction mechanisms but remain close to what they see as the most probable, that is the most empirically adequate, interpretations of their results. Quite probably they came across these quantities of helium in their apparatus simply by chance as they were developing their methods of helium detection; in any case, the underlying, essentially synthetic interpretation is more of a byproduct of the main line of experimental inquiry.

What makes them think the helium detected could stem from a transmutation? After all, both are chemists and in chemistry the elements are conventionally regarded as *natural kinds* par excellence and are both indestructible and preexisting. That means they are simply present in the world and cannot be manufactured by chemical means.[14] If one judges him according to his comments on the status of elements, Paneth here has contradicted himself in the sense that in his view no splitting of atoms or unmixing of isotopes can occur in chemical reactions (see

12 Cf., for example, Sun & Tomanek (1989).

13 With respect to the discovery of elements it is noteworthy that at this time these discoveries, and therefore probably the elements themselves, are being assigned to the chemical sciences (and consequently to the Nobel Prizes in chemistry) while their universally accepted systematic representation, the periodic table, is being converted to a form typical of physics.

14 The concept of "natural kind" in the philosophy of science is of course of recent origin although we may assume that most natural scientists long before the historical period under discussion here regarded the chemical elements (in Lavoisier's sense) in precisely the same way as is customary in modern discussions (cf. Chapter 7). In chemical thermodynamics only compound substances, that is chemical *compounds*, have an *enthalpy of formation* other than zero. The value of this enthalpy for elements is (according to the prevailing convention) zero.

quotation above). For helium to form from hydrogen, the hydrogen would have to give up its internal composition.

Element transformations are also discussed in the textbook on radioactivity cited above, for example, the "splitting" of nitrogen by α-rays (Hevesy & Paneth 1923, p. 131). This process creates so-called "H-rays," that is protons, and oxygen:

$$^{14}_{7}N + ^{4}_{2}He \rightarrow ^{17}_{8}O + ^{1}_{1}H \tag{4}$$

The H-rays are demonstrated by their tracks in cloud chambers (cf. Rutherford 1919). Interestingly, this textbook does not refer to the transformation of nitrogen to oxygen but to the creation of hydrogen. In terms of atomic mass, the creation of oxygen corresponds to a formation reaction whereas hydrogen is in a certain sense a waste product, a fragment. The reason for the preference for describing the hydrogen probably lies in the fact that the extremely small quantities of oxygen produced could not be detected with the method used.

Accompanying the question about the reaction mechanism of the postulated transmutation is the question about the energetic aspects of this process in which we could also include catalysis. In the original article, Paneth and Peters make no secret of the problem that enormous amounts of energy must be involved in the postulated transformation – especially in comparison with the usual heat phenomena in chemical reactions. Yet they make do with the assumption that in light of the minimal quantities of helium observed in the experiment these quantities of energy would hardly be distinguishable from other heat effects such as the adsorption energy released (Paneth & Peters 1926, p. 961). The Berlin group also performs experiments to examine the possibility of the release of radiation instead of heat, but the results are negative. However, they neglect one crucial aspect of catalytically controlled processes: When the "unification of hydrogen atoms" according to reaction (1) or an analogous mechanism can be accelerated by palladium then this should involve the rapid achievement of equilibrium of the process. According to the prevailing opinion (from 1926 until today), the catalyst decreases the barrier to activation of a chemical process. Yet if it actually does this, then the quantity of product must be detectable and abruptly increase because as a rule the process, once activated, continues to proceed by itself due to the reaction energy released.

This conflicts with the observed "course of the reaction" (which apparently is not one) and the extremely small "quantities of product."

So what went wrong with the Berlin group's interpretation? As was mentioned, the authors write in the original report that even in the case of contamination of the palladium from the helium content of the air, the additional hypothesis would be required that only hydrogen can release this helium. Nonetheless, they test the permeability of palladium for helium (with and without hydrogen) and conclude that hydrogen shows no effect in this experiment. The obvious idea (in hindsight) of testing all other equipment materials, such as the glass tubing, for comparable behavior does not initially occur to the authors. Only a short time later it becomes apparent that it actually is the glass that takes on helium which, especially at higher temperatures, is displaced by hydrogen (but not oxygen) and can thus be released. Paneth then publishes the correction (Paneth 1927a, 1927b), which in German bears seemingly the trivializing title *Neuere Versuche über die Verwandlung von Wasserstoff in Helium* [Recent Experiments on the Transformation of Hydrogen into Helium] and in which he concludes that the misinterpretation cannot call the essential metrological fundamentals into question. One can simplify the final argumentation in an implicative manner as follows:

Helium is only detected in the setting of the given detection sensitivity in the present experiment when

a) glass is used the vessel material and
b) this glass has come into contact with air containing helium and
c) helium from this air is adsorbed or absorbed and
d) helium is released upon contact with helium-free hydrogen.

Assumptions (a), (b), (c), and (d) represent *additional hypotheses* as have been described in the classic philosophy of science, for example by Carl Gustav Hempel (Hempel 1974, pp. 36 – 40). All four must apply, whereby it is easy to see assumption (d) is the one that posed the greatest problems for the Berlin group. However, one first has to hit upon (a)–(c). Once these additional hypotheses have been considered, the experimental decision – in a certain sense about an *experimentum crucis* (Hempel 1974, pp. 40 – 44)–between the following rival *test implications* (e) and (f) becomes rather unproblematic:

e) When helium is detected as hydrogen flows through hot palladium, then that helium is attributable to a transformation reaction of hydrogen.
f) When helium is detected as hydrogen flows through hot palladium, then that helium comes from the glass wall of the apparatus, which has been contaminated by air.

It is easy to verify that helium is absorbed by glass under the given experimental conditions: "Our method of detecting helium is sensitive enough to show that a glass tube that has been freed of its helium by heating it in hydrogen absorbs a detectable amount of neon-free helium from the atmosphere when exposed to the air for even one day." (Paneth 1927a, p. 379) After this it is only necessary to demonstrate that without the measured "blank value" no helium is indeed detectable.[15] The minimal amounts of helium would thus be measurable using unprepared glass vessels even without the palladium. Expressed somewhat more bluntly, the cause of the Berlin group's erroneous conclusion is essentially an overlooked blank value.[16]

It can be inferred from his correspondence that Paneth is troubled by the public dissemination of his interpretative error. He writes to his brother Otto on 16 October 1927,[17] "You mention my helium work in your last letter. As I do not know whether my second notice has achieved the same publicity in your area as did – very much against my will – the first, I am sending you enclosed the supplemental correction [...]. This "supplemental correction" is undoubtedly the article mentioned above that was submitted for publication in April 1927. It is also obvious from his correspondence with his coworker Kurt Peters that the latter had encouraged the publication of the article rather than regarding it with critical restraint. In this particularly lively correspondence during Paneth's time at Cornell University, Peters also reports doubts on the part of other scientists such as Otto Hahn and Lise Meitner.

This suggests the question of whether the retraction of the previously announced scientific sensation damaged the main author's reputation. It is known that after their period of working together at the Institute of Radium Research, the two friends Paneth and de Hevesy became competitors for professorships. In his efforts to find a successor to Georg Meyer as professor of physical chemistry in Freiburg, Heinrich Wieland (1877–1957) negotiated first with Paneth (who in his own words had nearly accepted[18]), then with Fajans, and when he too declined, with de Hevesy who then in 1926 accepted the professorship in Freiburg (Niese 2007). Paneth transferred from Hamburg to Berlin in 1922, where his helium research began. Measured by this episode in university politics, one can surmise that Paneth's general scientific reputation in the 1920s was at least not inferior to de Hevesy's. Paneth was nominated for the Nobel Prize a total of 34 times. The

---

15 In analytical chemistry, the unspecific measured value that is obtained under otherwise identical conditions without any analytes is known as a *blank value*. It must be carefully noted and subtracted from every further measurement. The background noise of measurement devices is a similar concept.

16 I expressly refer here to a classical logic of argumentation because Paneth argues in such an exemplary and transparent manner that one is struck by the thought of applying to it the classic methods of logical argumentation. It is apparent here that this process of interpretation has its advantages particularly in local contexts (that is close to the discussion of the empiricism of the particular sciences). Nonetheless, Paneth's own inclination to a logic-centered philosophy of science (such as that of Carnap) is quite limited.

17 Cf. the letter in the archives of the MPG (Sec. III, Repository 45).

18 See his letter to Wilhelm Schlenk (1879–1943) of 3 January 1921.

first time was in 1927–interestingly for the first article discussed here, together with Kurt Peters – and later for his research into alkyl radicals (see next section). De Hevesy received 31 nominations, beginning in 1924. Several times, once even by Paneth in 1929, he was nominated in conjunction with the x-ray spectrometric detection of the element hafnium. He received the prize in 1943.[19] Paneth never received a Nobel Prize although he was awarded many other honors and after a short time was assimilated into the academic world in England, initially in London, later as a professor at the University of Durham.[20]

I would like to conclude this section by emphasizing the following points. First, it is remarkable that in light of his conviction that the elements are indestructible Paneth even considers the train of thought discussed above. Second, the extreme care with which the experiments were conducted, the transparency of the interpretation (including the error discussion), and not least the archived correspondence, all show that this was no pseudoscience such as can be assumed for corresponding publications after 1989.[21] At least in Paneth's day, the transmutation of chemical elements at low temperatures ("cold fusion") was not thought to be excluded in principle.[22] And finally, it is noteworthy that the project discussed here pursues an unusual synthetic premise that transfers the manufacturing principle from substance chemistry to the field of nuclear reactions. A similar process occurs in what has now become the established research field of the super heavy elements (SHE), albeit with plausible empirical evidence. In any case, the "chemical" grasp at "physical" entities seemed to be actually possible for a brief moment in history. Even after his bitter scientific setback, Paneth did not give up the idea of transmutation and the important role that chemistry would play in studying it:

> It may well be that there are processes of transmutation going on with so little energy that the radio-physical and radio-chemical methods of detection cannot be applied; there is now hope that in some such cases ordinary chemistry will be able to discover the newly [sic] elements. (Paneth 1936a, p. 562)

---

19 Naturally, there is no causal relationship between the number of Nobel Prize nominations and the probability of winning the prize. The extremely influential American Gilbert Lewis (1875–1946) received 41 nominations without a prize whereas one (by The Svedberg) was sufficient for Harold Urey to be honored in chemistry (he received a further nomination in physics that same year).
20 Cf. Emeléus (1960).
21 Cf. the presentation of cold fusion (albeit not consistently historically accurate) by Huizenga (1993). Philosophical analyses on the subject of pseudoscience may be found in the collection by Pigliucci & Boudry (2013).
22 In a series of articles Paneth pursues transmutation in his own projects until well into the early phase of his English exile. See for example Paneth & Günther (1933; undertaken in Königsberg), Paneth & Loleit (1935) and Paneth & Fay (1935); and, as was cited previously, Paneth (1936a).

It is remarkable that Paneth not only published this statement, which from a current perspective is untenable, several years after his retraction but also several years after the bitter personal setback of his exile. In the year of the Olympic Games it had been three years since he had lost his professorship, although he continued his scientific work, for example, his publishing activities, without any outwardly discernible limitations.

9.3 Non-substance Chemical Entities: Alkyl Radicals

Toward the end of his time at the Chemical Institute of Berlin University (1922–1929) Paneth and his coworkers began to conduct research in another field while the helium experiments continued.[23] This new field of research also followed a synthetic thread insofar as it pertained to the characterization of new chemical entities.[24] It involved experiments to describe "free alkyls," i. e., research into "free radicals." After they had "found indications of the viability of free methyl" Friedrich Paneth and Wilhelm Hofeditz, in an exceptionally skillful experimental approach, attempted to detect "the effect of free methyl at a distance of several centimeters from the site of its occurrence" (Paneth & Hofeditz 1929, p. 1336). They started with lead tetramethyl $Pb(CH_3)_4$, which they obtained via a Grignard reaction and introduced into a hydrogen stream in small quantities at low temperatures. Figure 24 shows a schematic diagram of the relevant section of the apparatus:

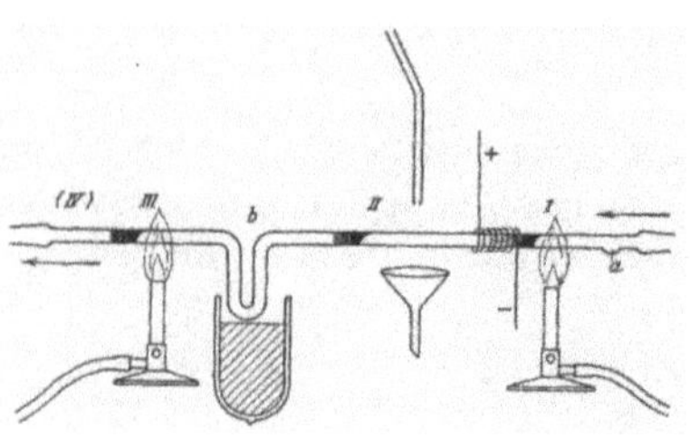

Fig. 24: Section of the apparatus for indirect detection of the methyl radical, from Paneth & Hofeditz (1929, p. 1337).

23 His interest in helium is basically motivated by cosmology and epistemology. Until the end of his life he poses questions about the age of the world and the possibility of answering such questions by chemical and analytical means.

24 Whereas the idea of element synthesis appeared by chance – and ultimately erroneously – Paneth worked systematically on the description of previously unknown substances in his studies of hybrids of a series of metals (such as bismuth, polonium, tin, lead, and germanium). He provides a comprehensive overview of these "volatile hybrids" in his Cornell lecture, see Paneth (1928, pp. 91–116).

This is a quartz tube about 60 cm long with an internal diameter of 0.5 cm. The hydrogen charged with lead tetramethyl flows in from the right. The apparatus is largely evacuated and the gas flow rate is set at 12–16 m/s. At point I the lead compound is thermally dissociated: $Pb(CH_3)_4 \rightarrow Pb + 4CH_3$. The lead precipitates out as a silvery "lead mirror." The same process is repeated at point II in a new tube. Then we come to the crucial point of the experiment:

> If one now precipitates a new lead pool at point I in the same manner, one can observe how the *old pool* at point II decreases during this process and, assuming the reagents are completely pure, quantitatively *disappears from the tube.* (Paneth & Hofeditz 1929, p. 1338; original author's emphasis)

The same phenomenon is also observed with highly pure nitrogen as a carrier gas. The authors specify the half-life of the methyl radical (in their experiments, the authors do not pursue the "decay" to ethane thought to be more probable at the time) at about 0.006 seconds. Their crucial conclusion is:

> Thus it can be seen from the previous experiments that *with the thermal dissociation of lead tetramethyl a gaseous medium arises* which can dissolve metallic lead and convert it into a compound similar if not identical to lead tetramethyl. This agent exists only briefly for its efficacy decreases greatly as the distance between the site of its occurrence and the reaction site increases. (Paneth & Hofeditz 1929, p. 1340; original author's emphasis)

Consequently, the next radical to be studied was ethyl, C_2H_5, for which Paneth and his coworkers received results similar to those of methyl.[25]

The alkyl findings were interpreted as a major breakthrough and soon made Paneth an expert on radical chemistry in the eyes of the scientific community.[26] The Cambridge physical chemist Thomas Lowry (1874–1936) emphasizes this as follows in his introductory address to a discussion conference of the Faraday Society he organized on the subject of "Free Radicals":

> Exceptional interest attaches to Paneth's isolation of methyl and ethyl radicals as products of decomposition of lead tetra-methyl and tetra-ethyl. Although these radicals may have a life of only about 0.006 seconds it has been possible to use them in synthetic work. (Lowry 1934, p. 9)

---

25 See Paneth & Lautsch (1931a, 1931b) as well as Paneth & Herzfeld (1931).

26 In the the archives of the MPG (Sec. III, Repository 45) there are indications of lectures in London (17 and 24 January 1935, 12 January 1939), Bristol (8 March 1935), Sheffield (15 March 1935), and Cambridge (21 August 1950).

Paneth's speeches and lectures continue to show his philosophical reflection on science, which is barely hinted at in his technical publications. In a handwritten outline for a lecture (presumably from 1950), he rephrases Paul Walden's classification conditions for the viability of radicals, which are relevant not only for chemistry, as questions:[27]

1. Can we isolate radicals and study their properties like those of other molecules?
2. Are they not stable in the free state, but do they play a role in chemical reactions?
3. Are they only mental conceptions, without reality in the outer world?

The classification that Walden had published a few years before the papers of Paneth's group is essentially applicable to all chemical species, especially those that are significantly less substance-related. Here are two examples that illustrate the limits of such a classification system: Carbonic acid (empirical formula H_2CO_3) does not exist as a substance, yet its derivatives populate our planet in enormous quantities (for example the Dolomite Alps consist primarily of calcium carbonate and magnesium carbonate). We cannot suitably describe carbonic acid as a substance and not at all as a chemical individual in František Wald's sense, yet we can describe it as a chemical entity in the wider sense. The "artificial" elements (super heavy elements or SHE) are another example. These do also not exist as substance objects but, if at all, then as small numbers of atomic nuclei from which one can wrestle a few "characteristics" only with extraordinary skill due to their short lifespan. In contrast to most of the other chemical entities we have called elements since Lavoisier, the SHE have no reference in the world of everyday life; in the words of Jens Soentgen: they do not occur.[28] Therefore I regard the term "chemical element" in these cases as not convincing or better: inapplicable.

In any case, it is obvious that the alkyl radicals of Paneth's working group belong in the second class of radicals: One can visualize them and demonstrate their existence but not synthesize them.[29] They do not have a substance character but can nonetheless have chemical influence.[30] Although Paneth very probably knew Walden's radical book in 1929, it cannot be confirmed by available sources. In

27 See the passage in Walden (1924, p. 3) cited in Chapter 2.

28 We learned from Wald, however, that all other entities we call elements rather have to be prepared as well.

29 We can assume that this is exactly what Lowry means when he uses the word "isolation."

30 See Paneth (1934). In this article for the meeting in Cambridge in question, Paneth reports on reactions of "methyl" and "ethyl" with antimony, arsenic, and bismuth, that is on the synthetic possibilities of the radicals.

any case, it is not mentioned in the radical publications. A bit later Paneth refers among other things to the epistemological significance of the knowledge about radicals in chemistry; I will discuss this in the next section.

9.4 The Kantian Echo: The Königsberg Lecture

In 1929 Paneth joined the Department of Chemistry in Königsberg where he became director of the chemical laboratory. His research interests remained helium and the radicals. On 12 July 1931 he delivered a public lecture before the Königsberger Gelehrten Gesellschaft in which he formulated the essence of his philosophy of chemistry. The lecture was published that same year in the writings of the Königsberger Gelehrten Gesellschaft (natural science class) under the title *Über die erkenntnistheoretische Stellung des chemischen Elementbegriffs.* Paneth devotes himself to the task of subjecting fundamental chemical concepts[31] to epistemological clarification and laments chemistry's lack of pervasive philosophical treatment:

> [...] between philosophical and physical writings there has long been an extremely fruitful interaction to which we in the field of chemistry have nearly nothing comparable. (Paneth 1931, p. 102)

Regarding the epistemological questions that were first posed in ancient times and remain relevant in chemistry to this day, he then states:

> I mean here primarily the question of in what sense one can assume the persistence of the elements in compounds. (Paneth 1931, p. 102)

In addition to "philosophical clarification" Paneth, in the same passage, expects the opportunity to assess the merits of the notion "that chemistry should and will dissolve into physics." He describes the relationship between the substances and the characteristics that can be perceived in them as follows:

---

31 Aside from the concepts "substance" and "element" discussed in the main text, he mentions as examples "qualitative character, valence, etc." in a parenthetical remark in the introduction (Paneth 1931, p. 101). – In a letter of 31 March 1932 to the philosopher Ernst Cassirer in Hamburg, he tells him that he wrote the manuscript for the Königsberg lecture three years previously after intensively studying the works of the German philosopher Eduard von Hartmann (1843–1906). He sends Cassirer the printed article as an enclosure (Paneth 1931).

> When one engages the problem of substance more resolutely with the weapons of critical epistemology, the fact is revealed, as has been stated from many sides, that not only may none of its sensory characteristics be assumed to exist in nature, but that it itself evaporates. (Paneth 1931, p. 104)

"Erkenntniskritik" defines the methodology of his examination. It leads, as with Kant, to the initial result that the intelligible world cannot be assumed to actually exist: Although chemistry, which concerns itself with the perceptible transformations of substances, hardly leaves the "ground of naive realism," it strives toward the pervasive theoretical understanding of its field of knowledge. In this process, the descriptions of the substance qualities from the world of phenomena are replaced by scientific concepts, and, according to Paneth, the substance concept gives way to expressions of the description of an underlying reality. The "process of evaporation" of the substance concept mentioned above found only one "point of retention very useful for the natural sciences" in Locke's differentiation of primary and secondary qualities. The result of this evaporation is the transition of chemical notions to atomism and the postulate that all substance characteristics are explainable based on the atomist theories.

> The subjective nature of the primary qualities that according to this view are supposedly still inherent in the things was demonstrated by the familiar idealistic argumentations (*Berkeley, Hume, Kant*). Indeed, even proponents of transcendental realism who are convinced of the existence of a nature outside of us that is knowable to a certain extent see themselves forced, if they take atomism to its logical conclusion, to declare substance as "nothing real, rather only a phenomenon in and for consciousness" that is brought forth by the interaction of forces lying in close proximity to one another. (Paneth 1931, p. 104; original author's emphasis[32])

With the expression "transcendental realism" Paneth explicitly refers to Eduard von Hartmann's position whose epistemology, Paneth believes, is the one that "possibly most correctly grasps the essence of natural science" (Paneth 1931, p. 103). Von Hartmann distinguishes three fundamental forms of epistemology: naive realism, transcendental idealism (Kant's position), and his own transcendental realism. In the introduction to *Kritische Grundlegung des transzendentalen Realismus* he makes the following statements about both naive realism and Berkeley's standpoint:

> Both equally lack the critical prudence that leads to genuine, namely transcendental idealism, i.e., one which is conscious of the fact that its ideas are not real things and initially and di-

32 Paneth quotes Eduard von Hartmann here (1907, p. 73).

> rectly can claim no transcendental reality if they are not indirectly granted such. (von Hartmann 1914, p. 4)

Von Hartmann describes the main difference between the two transcendental schools of thought as follows:

> [...] whereas the latter [transcendental idealism] denies the transcendental meaning, validity, and usability of forms of thinking and viewing, and thus the entire content of consciousness, the former [transcendental realism] proposes it; whereas according to the latter every proposition about the thing in itself (whether it is or is not) is declared impossible, the former seeks the epistemological significance of the contents of consciousness exclusively in the indirect insights that it provides about the things in themselves. (von Hartmann 1914, p. 8)

He feels it is possible to provide an "indirect proof" for his position by a *reductio ad absurdum* of the Kantian position. However, it remains to be discussed whether von Hartmann's characterization correctly portrays Kant's view.[33] Von Hartmann's transcendental realism does not refer to things or objects, that is empirical entities, but to the "transcendental cause" or the "basis for the phenomena"; it is thus a theoretical realism. In the Königsberg lecture Paneth emphasizes the presumed difference between Kantian idealism and von Hartmann's transcendental realism as well as his inclination to the latter, probably to keep himself as a natural scientist on the safe side with the realists. He wants the expression "transcendental" to be understood in a purely epistemological sense. This raises fundamental problems.

Access to scientifically interesting fields and thus to chemistry as well is empirical, that is it is gained through the world of the phenomena. This is a conviction that links the philosophers Kant and von Hartmann to the natural scientist Paneth. All three reject a naive entity realism, that is the belief in the "reality" of the environment accessible to the senses. All three propose or utilize the noumenal-phenomenal dichotomy of the world of knowledge proposed by Kant with the appropriate concepts as well as the epistemological attitude with the striving for knowledge about the transcendental world. The crucial difference of Kant's position is that even as sciences advance the noumenal world is still seen as not completely explained. Paneth agrees with von Hartmann in the assumption that the world is at least partially knowable.[34] Paneth also shares Kant's early skepticism

33 In the *Critique of Pure Reason*, we find a "refutation of idealism", in which Kant argues against the skepticism concerning the external world of Descartes and Berkeley, see Kant (1781, pp. 147–156, or B 274–279).

34 We can assume that Kant, too, felt scientific advancement was possible insofar as previously inaccessible phenomena or knowledge can be made available by work in the phenomenal

about the scientific nature of (pre-Lavoisier) chemistry due to its low content of mathematics. He states:

> If one accepts this definition, Kant is completely correct in not including chemistry among the sciences because it is by nature unmathematical; he saw this more correctly than later philosophers who expected it to merge with physics. (Paneth 1931, p. 106)

With this statement Paneth effectively portrays the matter of the difference between the sciences of physics and chemistry positively with respect to the latter: If one regards chemistry as dissimilar in nature to physics, then its theoretical reduction to physics is hardly possible. It is also striking that the ordering principle of the periodic table of the elements – disregarding for the moment the correction by Moseley – was found based on the "most primitive concept of substance" and the atomic physicists were astounded to find "that the analysts had already discovered almost all possible elements." Thus Paneth concludes:

> A "systematic art" that can achieve that is one we can arguably call a science. (Paneth 1931, p. 107)

Paneth disagrees with Kant's classification of the sciences. He finds it too narrow in general, and its concept of science "arguably not expedient either." Following Kant he distinguishes a world of phenomena and the possible subjective perceptions contained therein from a fundamental objective world in which the laws of nature apply. He repeatedly refers to this empirically inaccessible world in the Königsberg lecture as "transcendental," precisely in the Kantian sense of *going beyond experience*. Paneth's dualistic conviction and the significance of the "objective world" for scientific chemistry becomes particularly clear in the following passage:

> The numerous relationships between the compound forms of the chemical elements that are expressed in the natural system perhaps best make it clear that standing behind the "simple substances" with unchanging characteristics and "compounds," which like Proteus bring forth innumerable forms and yet are bound to rigid rules, are transcendental "basic substances" and that it is the task of chemistry to study the laws of this transcendental world. (Paneth 1931, pp. 118–119)

The actual natural laws of chemistry in Paneth's view must therefore be sought in the transcendental field, yet we only acquire knowledge about them in a round-

realm. The difference between Kant and von Hartmann is essentially not as great as the latter portrays it.

about way by work in the empirical field. It seems conceivable that Paneth was inspired to use the term "basic substance" by having read Kant.[35] The latter uses this expression in several places, for example also with reference to contemporary notions of acids and bases:

> It was considered a great step when chemists [Scheidekünstler] were able to reduce all salts to two main genera – acids and alkalis; and they regard this difference as itself a mere variety, or different manifestation of one and the same fundamental material [Grundstoff]. (Kant 1781, pp. 365–366, or A 652–653)

The function of the expression "basic substance" is here very similar to that which Paneth envisions for the "transcendental" elements.Paneth thus sees metachemical premises as unavoidable in order to form a complete picture of the sciences of substances. Questions such as this one must be answered (Paneth 1931, p. 116): "What sense does it make to say that the unchanged element sulfur is contained even in its compounds, for example gaseous, colorless, pungently smelling sulfur dioxide?" Paneth answers this question on the basis of a dualistic separation of the element concept (see Figure 23). What persists in all substance transformations, ignoring for a moment radioactive decay, are the atomic nuclei, that is what is listed in the periodic table with a symbol and an atomic number; Paneth calls this the "basic substance." And this basic substance, which is actually not a substance at all, is not directly accessible to experience.

> *The fundamental law in chemistry that the elements remain intact in the compounds refers to basic substances without qualities*; in any other sense the claim is incomprehensible. Referring to Ostwald and understanding the statement only in the sense that there is the "possibility" of "recovering the element from its compounds in unchanged quantity" is not a way out, for if nothing else is really being claimed, then one would not be able to speak of "persisting in the compounds." (Paneth 1931, p. 118; original author's emphasis)

Paneth categorically rejects Ostwald's views, which are based exclusively on real-world ("simple") substances, because at least with respect to the aspects under discussion here they are strictly operationalist and sensualist and thus descriptive in nature. Paneth, surprisingly, feels that the premise that all substances are suitably characterized only by a neutral description of their specific properties which are

---

35 Already Dmitrij Mendeleev (1834–1907) had very similar ideas and it is known that Paneth was familiar with his relevant writings, at least in English and German translation; cf. Paneth (1936b). However, it is not my interest here to pursue the discussion of priority. See Boyce (2019) on this topic.

to be sought in the realm of experience is too narrow. At this point he again emphasizes that the sciences remain impotent without Kant's premise:

> There is likely no other example that shows so convincingly that the simplest and most familiar laws of natural science have validity only with the tacit or express assumption of a transcendental world. (Paneth 1931, p. 118)

At least two things are questionable here. First, it remains to be clarified how the application of Kant's epistemological framework can be squared with the "Lavoisier" definition of element mentioned here, which Paneth emphasized earlier and initially favored in his Königsberg lecture.[36] Second, the problem mentioned above arises as to how chemistry is to be defined if the substance concept on which "the entire theoretical edifice of chemistry rests" has "evaporated." Paneth argues for the operational notion of element as opposed to the four-element doctrine by noting that the number of elements is determined a priori and not speculatively.

However, in his judgment this does not fundamentally (metachemically) change anything:

> [...] the essential thing is indeed the continued assumption that the simple substance in some form is present latently, i. e., without revealing the characteristics it has in pure form, in the composite ones and explains their behavior. (Paneth 1931, p. 114)[37]

A number of substances – Paneth mentions sodium, fluorine, and radium as examples – were assumed to be basic substances long before they could be assembled operationally because experience with appropriate empirical specimens suggested this. Basically, the dichotomy of the element concept that he specifies in greater detail was already in use earlier, albeit implicitly. The relationship between the world of phenomena and the transcendental world that is so crucial to the study of elementary substances also plays an exceptional role in the evaluation of the significance of the chemical substance concept. The transcendent basic substance, which one therefore must view as "overflying" and not "indigenous,"[38] is now allocated to new non-substance structures, the molecules and atoms. These "are admittedly no longer solid spheres; the space ascribed to them is filled only

36 Cf. the discussion of his "practical" and "theoretical" element definitions in the relevant section above.

37 Note that the "simple substance" mentioned here is equivalent to "element" as Lavoisier understands it but Paneth's proposal refers to the "basic substance."

38 I borrow the terms "überfliegend" and "einheimisch" (transcendent/immanent) from Kant (1781/1980, p. 564 or A 643). English translations usually neglect this pair of illuminating notions.

with electrical forces" (Paneth 1931, p. 105). The old materialism has thus exhausted its usefulness:

> Now one could understand protons and electrons as the last remains of the old concept of substance had not modern quantum mechanics forced us to assume that the movements of these particles do not occur in the three-dimensional space we can imagine but in spaces of far higher dimensions. With this succumbs the last possibility of clinging to our familiar concept of substance in some sense; all clarity has been sacrificed and predictability has simultaneously been raised to a height hitherto not attained. (Paneth 1931, p. 105)

Elementary particles such as protons and electrons are for everyday experience, at least the usual realm of chemical work and experience, non-empirical (theoretical) entities. In the Kantian sense they belong in the transcendental realm of ideas. With the abandonment of the concept of substance described here, which despite all abstraction – see Peter Hacker's analysis discussed above – can indisputably be understood as belonging to the intelligible world, it becomes clear that "basic substances" cannot be substances in the conventional sense: They are without qualities.

The "evaporation" of the substance concept Paneth mentions is also applicable to other chemical species. If we examine the chemical radicals, we see that they, too, lie in the no man's land between empirical and theoretical observations. Paneth feels that the requirement of demonstrability of this chemical species is unfounded – we learn that from the history of elements such as sodium, fluorine, and radium – for its correlate in the world of phenomena does not need to be isolated to prove its existence because:

> The "radicals" of organic chemistry for example exist almost without exception only in the transcendental world [...]. The radicals are even a still more convincing example of the transcendental character of the chemical laws of persistence because – in contrast to the elements, of which the majority can be easily demonstrated as simple substances – the old proofs of the independent, perceptible existence of the free radicals were all wrong and were soon disproved, and despite this the acceptance of the radicals was completely justified. (Paneth 1931, p. 122)

According to Paneth we can ascribe a dual character to the radicals as well, quasi an existence without a substance correlate. As has been implied, the strict interpretation of the view that the transcendental world has priority over the phenomenal world in the sciences would have fundamental effects on chemistry. If the substance concept were abandoned entirely, then the above-mentioned definition from a well-known dictionary of chemistry would be correct:

> Chemistry is the science that concerns itself with the causes and effects of the shedding, acquisition, or distribution of electrons between atoms or molecules and with the relationships between the energy levels of such electrons within the atoms or molecules. (Römpp 1975, p. 548)

Chemistry would have been assimilated into atomic physics long ago were this characterization true. Naturally, we perceive changes in substances only in the world of phenomena, that is based on “simple substances”; yet here we are speaking exclusively about entities that cannot be directly observed. Is such a definition of chemistry really fitting? I think not. First, a good definition should be characterized by conceptual economy, meaning that it should avoid words that themselves require explanation. Electrons; atoms; molecules; the shedding, acquisition, or distribution of electrons; and energy levels: Even just the list of concepts that come from the particle-centric vocabulary of modern chemistry and which according to Kant and Paneth can all be deemed to belong to the transcendental world is remarkable in this regard. Second and closely related to the first aspect, the definition of a science should be able to say something to the inclined but uninformed reader, for example, a schoolchild. This can hardly be the case when relationships from the current state of this science are used. And third, the definition should be clear in epistemological and methodological terms. To begin a characterization with terms from the current theoretical findings of the science reverses the methodological order that can legitimately be expected. Indeed, chemistry does not concern itself with matter, atoms, or elementary particles. It acknowledges them and uses corresponding concepts in the sense of theoretical entities but only as long as they prove useful in scientific work. The “evaporation” of the substance concept may not be, and is not, confused with the “dissolution” of the concept that has already occurred as long as one is speaking of actual chemistry and not its theoretical decal. From my perspective, the “substance concept on which the entire theoretical edifice of chemistry rests and which is also fundamental to the element concept that is of interest specifically to us here as well” should remain until further notice. In a final summary of his Königsberg lecture, Paneth states:

> Defining and explaining the peculiar resulting epistemological doubling of the element concept, which, as has been shown, can be understood in the naive-realistic sense of “simple substance,” yet in the sense of “basic substance” must be understood transcendentally, and emphasizing the close relationship of the concept of basic substance with the “metaphysical” principles of the Aristotelians and alchemists was the intent of the present study. (Paneth 1931, p. 125)

Throughout history, what we have called a (chemical) element has been the result of conventions,[39] and these can change. As regards Paneth, the crucial question remains as to the extent to which we can be certain of the stability of the inventory of the noumenon. Paneth's writings in particular have shown us that the limits of knowledge, or at least what one once assumed them to be, can be shifted. Is then the existence of protons and neutrons in the atomic nucleus really transcendent, that is not demonstrable or knowable in any way? Obviously not. In physics, for example, experimental methods have been developed to detect the elementary particles and determine the atomic number. What Paneth means by "transcendent" is thus better interpreted as "not directly discernible with the methods of substance chemistry". Whatever is experimentally inaccessible in chemistry is not necessarily inaccessible in physics.

9.5 Conclusions

In his study of 1916, basically a methodological one, Friedrich Paneth shows that there are different ways to characterize chemical elements, and he doubles the element concept by referring to a "practical" perspective and a "theoretical" one. Although he greatly emphasizes the "practical" perspective for some time, he gives higher priority to the "theoretical" sense, ontological priority as it were. The relevant international commissions (and finally the IUPAC as well) have since propagated only this physicalist sense, thus contributing to the increasingly prevailing microphysical image of the science of substances.

In the Königsberg lecture of 1931, Paneth interprets his element doubling against the background of Kant's epistemology. The "theoretical" interpretation of the elements merges into the concept of the "basic substance," which is without qualities (because it is noumenal), and means the atomic number; the substance specimens accessible to practical chemical manipulation that can no longer be separated are now, according to the record (Lavoisier), referred to as "simple substances."

What is important yet often overlooked is the fact that the concept of basic substance does not require the image of the complete atom: A persistence law can be formulated for the chemical realm of experience only for the atomic nuclei, although they are not observable and, as Paneth himself discovered to his chagrin, not chemically manipulable. Paneth's concept of element is in this sense non-atomist. This correlates with the modern understanding that the outer electrons in an

39 I have argued for this thesis in Ruthenberg (2020c).

atom are what determines its chemical behavior and with the thinking of specialists in super heavy element research. However, it also correlates with the fact that developments in particle physics have hardly any benefit for chemistry.

The "shifts" between previously inaccessible (transcendental, noumenal) and now empirically accessible phenomena or entities that can arise from scientific progress in certain fields of physics (such as high-energy, particle, and quantum physics) as well as in its own fields (analysis and synthesis of substances) are not events that chemistry must necessarily watch helplessly and leave to the supposedly more fundamental science of physics. Chemistry itself has experienced a number of cases that one can regard as a *revelation* of regularities or entities, as Paneth – whose own research invariably involved synthetic aspects – showed exceedingly skillfully in his research on radicals.

Whether a doubling of the element concept is tenable, however convincing it may appear at first glance, remains open to question. After all, even the knowledge of the atomic nuclei, for example by detection of the atomic number, is empirically grounded. And something that is measured can no longer be regarded as transcendental because for this measured thing the shift mentioned above has already occurred. I feel it is better to refer here to a pluralism of perspectives and to leave it to the respective scientific and philosophical discussions to decide which procedure one would like to call "chemical" and which "physical."

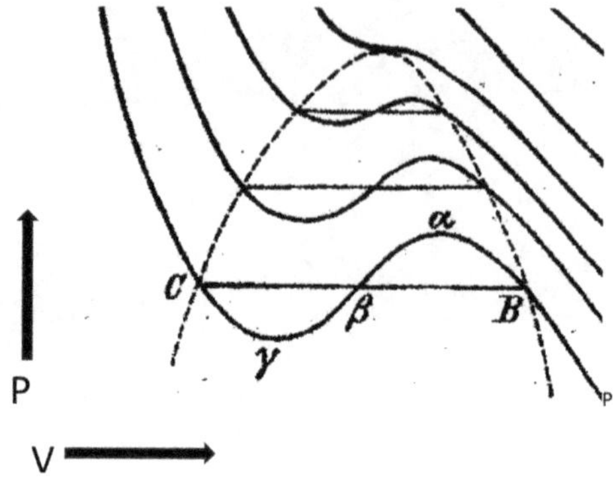

Fig. 25: The van der Waals curves in a pV-diagram (pressure vs. volume), an example of epistemological underdetermination of data: There are various interpretations for the same empirical situation. The area under the dashed line cannot be realized. From Ostwald (1890b, p. 88). See the following text for details.

10 Ontological Underdetermination

In this chapter I will examine a topic that has not yet been given any attention in the conventional physics-centric philosophy of science. It concerns an important aspect of the knowledge of the nature of chemical substances, namely their historicity, or rather a particular aspect of this historicity. As we have seen, the substance concept is derived from certain features of things by abstraction, that is by the conscious, focused disregard of certain aspects of the objects of our surroundings. Ida Freund (1863–1915), the first female university instructor of chemistry in Great Britain (Cambridge), describes chemistry as follows:

> Chemistry deals with the properties which belong to certain kinds of matter and not to others, which characterise one kind of matter and differentiate it from all other matter. (Freund 1904, p. 2)

As we have seen, it is the specific properties that are of crucial interest in the chemical sciences, not the question of the ultimate composition of matter.[1] The latter is the task of certain areas of physics, which at least partially explains its enormous philosophical attractiveness. Substances can be characterized in different ways.[2] Take, for example, the precious metal gold which occurs in the upper layer of the Earth's crust in pure form and therefore was known to humans very early, in the 6th millennium BCE at the latest. The fact that it occurs alone is an important piece of the knowledge about this special substance. This was soon joined by other pieces of knowledge such as fusibility, density, and malleability. Its insolubility in the liquids known at the time is a substance characteristic (or disposition) that belongs to the canon or web of characterization of this substance and naturally contributed to its being called a noble metal. Whereas once nitric acid was known one could remove gold's accompanying substances or "impurities" like silver from the mixture (the alloy), gold itself is soluble only in what is known as aqua regia. This, as we say today, is a mixture of hydrochloric acid and nitric acid and was probably first manufactured and used in the 8th century CE, presumably initially by chance. Since the first synthesis of this solvent, gold in our eyes – and that is the only perspective we know – has had a different "face." In the 20th century the atomic number joined this web of knowledge as a further character-

1 Here we may understand "matter" as "material," that is as a "weighable, extensive substance" (German: Stoff).

2 "[...] pure chemical substances are the relatively stable products of chemical analysis and synthesis: nodes in a network of chemical reactions." (van Brakel 1997, p. 253). See also the discussion in the final chapter.

https://doi.org/10.1515/9783111401041-013

istic feature and specified the substance's place in the periodic table of the elements. It is obvious that knowledge about an epistemic object increases with advances in the disciplines concerned with that object. Naturally, we may not assume that merely because we have discovered something surprising about a substance, like the solubility of gold, it is now something completely different than before. Gold is defined with sufficient precision and certainty by the web of fragments of knowledge mentioned. Ontologically it has remained the same for centuries although epistemologically it has been described differently at different times. In any case, regardless of whether we would like to regard certain characteristics or measurement results as essential or are satisfied with a fundamental perspective *based on description*, the history of a chemical substance can never be recounted completely. What interests me in this context is the term "underdetermination" used in the philosophy of science and its application to relationships in the science of substances. I will begin by discussing the conventional interpretation of this term.

10.1 The Customary Concept

The concept of underdetermination has been discussed for a long time in the philosophy of science and that discussion continues today.[3] To begin with, we can establish that this does not mean the lack of a conclusive theoretical description of a phenomenon, which in a certain sense is of trivial significance. Underdetermination refers to scientific theories and their relationship to empirical findings. The American philosopher of science Larry Laudan offers two versions of a definition of underdetermination:

> For any finite body of evidence, there are indefinitely many mutually contrary theories, each of which logically entails that evidence. (Laudan 2013, p. 290)

And:

> Any theory can be reconciled with any recalcitrant evidence by making suitable adjustments in our other assumptions of nature. (Laudan 2013, p. 295)

With reference to those philosophers he feels are relevant, Laudan names his first version the Hume version, the second the Quine version. Both are concerned with the various possibilities of scientific description of what is otherwise a solid stock of experiences (observations, measurements). The presence of multiple versions of

3 Cf. the overview in Ladyman (2002, pp. 162–195).

a valid scientific explanation seems to exclude an "experimentum crucis," that is a decisive experiment such as that envisioned by Francis Bacon (the so-called Duhem-Quine thesis).

Simple examples of underdetermination come from the mathematical interpretation of data sets. Curve fitting can assume various forms: A very few measuring points might initially suggest a linear relationship (under conventional aesthetic assumptions) that in reality may not be present, at least not over the full ranges of measurement. Extreme deviations from a linear interpretation (or discontinuities) are conceivable. A well known historical episode relevant to chemistry belongs in this class (curve fitting) of underdetermination: the so-called van der Waals isotherms. In his doctoral dissertation at the University of Leiden, the Dutch theoretical physicist Johannes Diderik van der Waals (1837–1923) advocated the theory of continuity between the liquid and gaseous states (van der Waals 1988). Based on mathematical and graphical modeling of known empirical data, he arrived in the course of this paper at the familiar state equation $(P + a/v^2)(v - b) = RT$ that bears his name to this day and significantly modifies Boyle's law of gases $PV = RT$.[4] The isotherms of van der Waals' theory are shown in Figure 25.

It was only later that the constants a and b received ad hoc physical and chemical explanations whose hypothetical content did not prevent many researchers from regarding the van der Waals equation as a realistic description, that is as an accurate representation of reality. This is obliviously illogical as the van der Waals isotherms, shown in the figure as sinus-like lines under the dashed region, cannot be represented empirically *in any case* and therefore do not "refer" to anything. Wilhelm Ostwald, whose *Grundriss der allgemeinen Chemie* initially (1890) included such a figure, at first discusses this theory rather impartially although he is also skeptical of whether the points α, β, and γ can be realized experimentally. In later editions of the same book, he dispenses with the figure entirely and mentions "strange conclusions" that result from the equation. At the same place he concludes:

> In this equation everything that depends on the specific nature of the substance has disappeared, and therefore it claims general validity like the equation of the perfect gases. The state equation of all substances would have to be identical if one expresses pressure, temperature, and volume as fractions of their critical values […]. One can […] regard this "law of consistent states" as a useful guide for a general overview yet not as a strict law of nature. (Ostwald 1917, p. 79)

4 The designations P, T, and V refer to the state variables pressure, temperature, and volume; R designates the general gas constant, and a and b are what are known as the "van der Waals constants." The "critical" values correlate with curve no. 4 above.

In the meantime (1910) van der Waals had received the Nobel Prize in physics for his creative handling of underdetermination, which was well received because it was soon to be compatible with the atomic hypothesis.

Hasok Chang discusses another example from the history of chemistry of the "underdetermination of theory by evidence." He identifies five main conceptual threads in the treatment of atomism in the 19th century: the reference to atomic weight, Berzelius' dualistic concept of electrochemistry, the Avogadro program of assuming volumes as measurable characteristics of atoms, the classificatory approach of organic chemistry with the type theory among others, and finally the geometric and structural orientation of certain researchers. These different orientations were actually pursued simultaneously throughout certain phases. It was only after a number of further experiences that the chemists – or at least some of them – were able to anticipate the unification of certain parts of what had previously been contradictory concepts.

> So they began to think that an oxygen atom really did bind two hydrogen atoms together in a water molecule, that a carbon atom held four hydrogen atoms to make a molecule of marsh gas (methane), and so on […]. With the synthesis of these three systems [the last three of the five main threads mentioned], nearly everyone came to agree that water was H_2O by the 1860s. (Chang 2012, pp. 150–151)

Underdetermination can, as we have seen, be interpreted in formal mathematical terms or historically.[5] The problems Chang discusses belong to the historical characterization. It becomes clear that this form of underdetermination is closely related to pluralism: Several partially contradictory views compete for the "best" explanation. What is important is that this competition is only partially resolved with the "confirmation" of the result because each original approach gives up or restructures its contents as the discourse progresses. It is not the case that one of the competitors "is right" and the others are discarded. Certain parts of the original main threads continue to retain their significance, such as Berzelius' electrochemical dualism. Chang criticizes both those who reject and resist underdetermination on principle because they see the sciences as striving for an ultimate truth as well as those who celebrate the coexistence of different uncertain theoretical approaches. He summarizes the section *From Underdetermination to Pluralism* as follows:

5 Wolfgang Pietsch (2012) refers to the Quine interpretation of underdetermination as "logical" and the Duhem approach as "historical" and argues for the latter.

> Scientists develop various systems of practice containing different theories, which are suited for the achievements of different epistemic aims. In this way the great achievements of science come from cultivating underdetermination, not by getting rid of it. (Chang 2012, p. 151)

Merely because the history of a chemical substance is never recounted completely, the search for the sole correct scientific description or explanation of its "nature" will have no prospect of success. Yet it is a different aspect that particularly interests me: Underdetermination, whose presence I feel cannot be denied, is implicitly or explicitly projected onto an ontologically unchanging frame of reference. Within it, the assumption prevails that the epistemic object in question is stable and constant. I call this conventional kind of underdetermination "epistemological," and I ask myself whether this kind alone is sufficient to satisfactorily characterize the chemical range of knowledge.

10.2 The Ontological Version

The assumption of a stable frame of reference does not apply to chemical substances.[6] Due primarily to human action, new substances are brought forth almost daily, intentionally and unintentionally. These substances often do not leave the laboratory. The example of aqua regia coming into contact with gold has shown that completely new phenomena can occur when previously separate substances encounter one another:

> Beyond academic interest, the production of nonknowledge by the synthesis of new substances is of general concern if the new substances leave, through whatever door, the laboratories and become part of our material environment. (Schummer 2001, p. 111)

The underdetermination arising from a shift in the stock of epistemic objects, in the stricter sense from an essential change in the stock of substance, and falling within the purview of the chemical sciences is what I refer to as "ontological underdetermination."[7] The ignorance (nonknowledge) regarding this that Schummer mentions is fundamentally different from the ignorance of the outcome of a game of handball in which the rules of the game and the general framework are predetermined. It is much closer to the ignorance of what game will or can be played.

---

6 Even the number of atoms of a certain element in a section of the universe, for example, the Earth, is known not to be constant.

7 I understand the adjective "ontological" to mean the occurrence or presence, essentially the stock, of substance-like objects in the observed system in the most general sense, cf. Meixner (1994). See also the discussion of the stock of substance in Soentgen (1997, pp. 90–98).

The introduction of new substances in the chemical space in question, that is the joining of substances previously isolated from each other, creates a fundamentally new framework for scientific processing. Predicting the characteristics of substances is therefore fundamentally limited in scope, especially when we look beyond local chemical spaces.

10.3 Chemical Substances between Knowledge and Ignorance

I have already mentioned gold above as a first typical example of chemical underdetermination. Whereas the presumably accidental mixing of gold with aqua regia did not change the actual nature of gold, it fundamentally changed our picture of that metal. At any rate, this is not a case of conventional epistemological underdetermination in which it is simply a matter of different theoretical interpretations of an empirical situation assumed to be constant. I will discuss other interesting cases in this section.

10.3.1 Thalidomide

This agent is well known in the German-speaking countries by the name of *Contergan*. Here is a brief outline of thalidomide's chronology of development:[8]

(1946) The company Chemie Grünenthal is founded in Stolberg, in the West German state of North Rhine-Wesphalia. Initially it primarily manufactured antibiotics. (1954) A patent application was made for the manufacturing process for "K 17" (the company's internal testing designation for thalidomide, which referred to the preparer Wilhelm Kunz who together with his coworker Keller had been hired that year and manufactured a total of 18 preparations). The process was patented and then licensed in 1960. (1955) Ciba introduces the competing preparation Doriden. K 17 underwent its first clinical trials. In April Dr. Jung of the Cologne University Medical Center reported his satisfaction with the agent and on 9 May he informed the director of Grünenthal's research department, Dr. Heinrich Mückter, of his continued good observations and stated: "The preparation seems to me to be ready for the market." On 16 December Grünenthal hosted the "K 17 Symposium" and presented pharmacological and toxicological results to selected invited ex-

8 This summary is based especially on Sjöström & Nilsson (1972), Knightley (1979), Kirk (1999), and Lenhart-Schramm (2016). For an ethical classification of the relationships and further literature see Ruthenberg (2016b).

perts. The experts mentioned side effects such as stupor, nausea, dizziness, headache, and hangover as well as itching and rash especially at higher and longer-term doses, but the manufacturer failed to conclusively resolve these issues. (1956) The only scientific publication from Grünenthal (Kunz et al. 1956) and the first clinical reports, circulated by Grünenthal, were published. On 11 July the Ministry of the Interior of the state of North Rhine-Westphalia approved the drug based on these reports. The attempt to sell a license to the American company Smith, Kline & French was unsuccessful because animal experiments conducted by that company failed to confirm the pharmacological effect claimed by Grünenthal. In November a field trial was conducted in and around Hamburg with the cold medication Grippex containing thalidomide (which has no known anti-inflammatory or antibiotic effect). (1957) Market launch of Contergan on 1 October (25 mg per tablet; Contergan forte: 100 mg per tablet). By this time at least 360 cases of side effects had been reported to Grünenthal. (1959) Neurologist Dr. Voss addressed a query to Grünenthal, asking to what extent long-term use of Contergan can lead to neurological damage. (1960) On 26 May after long hesitation, Grünenthal submitted the application documents to have the drug declared a prescription medication. (1961) The pediatrician and geneticist Dr. Widukind Lenz telephoned Mückter to convince him of the cause of congenital defects of the extremities repeatedly observed in children, namely the action of Contergan. Mückter attempted to pacify him. On 26 November an article was published in the German newspaper *Welt am Sonntag* at Lenz's urging, and on 27 November all forms of Contergan were taken off the market. (1961) The Australian gynecologist Dr. William McBride, who himself had prescribed thalidomide to pregnant patients in the first trimester, recognized the embryopathic effect and on 16 December published a letter to the editors of the medical journal *The Lancet*. Since then he has been regarded especially by the English-speaking medical community as the "discoverer," or at least one of them, of the embryopathic disposition of thalidomide. (1962) The US Food and Drug Administration (FDA) employee Dr. Frances Kelsey, initially without knowledge of the drug's embryopathic effect, ensured that the medication Kevadon with the active ingredient thalidomide was not approved for the American market. She based her decision on the incomplete and unconvincing documents presented by Grünenthal's American partner Richardson-Merell, documents based primarily on the studies circulated by Grünenthal. Kelsey resisted the significant and not exclusively professional pressure from this company with the result that the application was withdrawn in March. On 7 August she received the President's Award for Distinguished Federal Service from President Kennedy. (1964) Dr. Jacob Sheskin, director of a Jerusalem hospital, treated a man suffering from terminal leprosy with Contergan to allow him to sleep peacefully. Long since officially withdrawn from the market, the medication was from the hospital pharmacy's unused residual stock. This act of desperation was rewarded not only with the desired sleep but also with the remission of leprosy symptoms. This was the beginning of thalidomide's triumphant success as a medication for leprosy. (1968) In May criminal proceedings were opened against several official representatives of the Grünenthal company (the "Alsdorf trial"). (1970) The trial ended without criminal convictions. On 10 April a plea bargain was reached with Grünenthal paying 100 million marks. (1972) On 31 October the West German foundation *Stiftung Hilfswerk für behinderte Kinder* was established (since 2005 known as *Conterganstiftung für behinderte Menschen*). Grünenthal's contribution was supplemented by over three times that amount in tax money. (1976) The new Pharmaceutical Act came into force in West Germany. (1981) The AIDS crisis began. (1983) Sheskin and Mückter jointly published an article about the treatment of leprosy with thalidomide. (1993) Dr. Gilla Kaplan found out that thalidomide suppresses the activation of HIV. (1998) The Celgene company received FDA approval for Thalomid (thalidomide). Erythrema nodosum leprosum, a complication of leprosy, is specified as the indication. (2007) Lenalidomide was approved (like thalidomide it is an immunomodulatory drug of the structural group of imids).

Hardly anyone in Germany over the age of 60 will have forgotten the *Aktion Sorgenkind*, which today lives on in the form of a popular lottery known as *Aktion Mensch*. *Aktion Sorgenkind* came into being in 1964 as a reaction to the side effects of thalidomide. This agent was first synthesized in the mid-1950s in the laboratory of the Grünenthal company, at least according to the official version of the persons involved.[9] Surprisingly, the persons involved did not reveal to the readers of the only professional article their laboratory published on this topic (Kunz et al. 1956) how this synthesis was actually performed. However, this article claimed that thalidomide can be used very effectively as a sedative. This claim was based on the comparison of the effects of various preparations on groups of mice observed, according to the authors, in what is known as the "shivering cage experiment." Briefly summarized, the experiment looks like this: Eight male mice from the same dosage group are each placed in the same cage and their movements are registered by means of the electrolytic formation of hydrogen whose volume is used as a measure of their activity. The problem with this approach is that it is not possible to perform a clean calibration as not all the animals in any one group of mice are asleep at the same time. In a table in this astounding publication, thalidomide hardly surprisingly produced the best test results. The sensation is that the sedative effect that could not be established with certainty in the mice in

9 Regarding the early history of thalidomide and an ethical discussion of the relationships see Ruthenberg (2016b) and the literature listed there.

this experiment (which later led to US companies not licensing the manufacturing process) is paired in this table with a nontoxicity that is nothing less than fantastic. A value of >5000 mg/kg is specified compared with the competitor Luminal (phenobarbital): 300 mg/kg. The value assumed today, albeit in rats, is 113 mg/ kg when administered orally.

It remains a mystery to today's reader of these sources how the Grünenthal employees could actually have known about the sedative effect of their product. Published clinical studies of this do not exist. The fact that knowledge of this effect had been present early on is demonstrated by the ironically tragic case of the daughter of a company employee who was born without ears at Christmas 1956, one year before the medication's market launch. The employee had given the preparation to his wife as a sedative in an early phase of her pregnancy, that is in the spring of 1956. This fateful side effect (no others are known for this case) is in any case proof of the fact that the substance in question had already left the laboratory if not yet officially.

Many people (estimates are even as high as 10,000) have often had to lead a limited life of suffering with physical disabilities due to side effects that were not known at the time of the market launch.[10] In addition to this, one must assume that there is a much higher number of unrecorded stillbirths caused by thalidomide.

Who bears responsibility for what reason, when, how, and naturally to whom in matters of new chemical substances such as thalidomide? First, we should rid ourselves of the erroneous notion that a chemical substance has only good or bad characteristics. On the contrary Paracelsus' dictum still applies that it is the effective dose of a substance that determines its effect – desirable or undesirable – on people.

Let us have a look at who we can regard as agents. They initially include those persons who find a synthesis path and those persons (usually the same) who are the first to demonstrate it in a laboratory. I refer to all of these laboratory experts collectively as *synthesizers*.[11] *Producers* are those who manufacture the substance in question in larger quantities. *Testers* study the substance's applicability for achieving the desired effect (pharmacologists) as well as its safety and adequacy in the given legal framework (toxicologists affiliated with the producers and those affili-

10 None of the published clinical studies refer to pregnant women and in none are there any substantial indications of side effects; see Ruthenberg (2016b) and the literature listed there.

11 Substances that are formed by chance due to unintentionally created conditions and are then "discovered" analytically represent a difficult yet not uncommon special case, for example, intermediates in combustion processes, the Buckminster Fullerenes, or the so-called Gomberg radical (cf. Ruthenberg 2015).

ated with legislative and regulatory bodies. *Distributors* are those agents who bring the substance into circulation, as a rule, pharmacists and physicians. Then there are the *consumers* who independently and actively use a medication or receive it passively as patients in the narrower sense. Last but not least there are those at the end of the utilization chain who assume a cleanup function and whom we may call *disposers*. Each one of these groups of agents must be given consideration in a careful, balanced ethical examination. In order to be able to identify the individual areas of responsibility in the specific case of a drug like thalidomide, I find it helpful to use four of the principles or precepts that are central to modern medical deontology as a framework for evaluation:[12] (i) Respect for autonomy (principle of consent): The patient's wishes and ability to decide and act are to be respected and promoted; (ii) the precept of avoiding harm (*non-maleficence*): injury and suffering are to be avoided or prevented as a matter of principle; (iii) the precept of justice: the agents are enjoined to act justly in performing medically relevant actions, for example to pursue fairness and balance in the distribution of benefits, risks, and costs.

When we examine the early history of the case of Contergan, we cannot avoid concluding that poor science was performed here. This evaluation pertains initially to the chemical-synthetic side: The path to the manufacture of the drug is not even hinted at, and we learn nothing of its stereoisomerism (note the R-(+) enantiomer in the following structure).

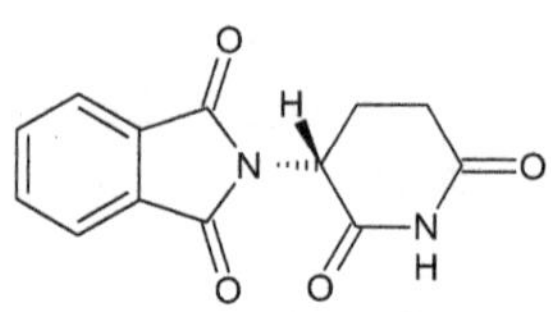

In this so-called "keto form" the marked hydrogen atom points upward and the phthalimide group (left) points downward. In the mirror-image form, the S enantiomer, this is reversed. The manufacturing process creates a racemate, that is a mixture of both forms.[13] After the noxious effects had become known, experiments were conducted in the 1970s on particularly sensitive rabbits (White New Zealanders) to determine to what extent the enantiomers react differently. For a while the rumor circulated that the R form was the "good" sedative agent and the S form was the "bad" teratogenic form. However, this line of inquiry cannot

12 Cf. Beauchamp & Childress (2009).

13 It is not known whether thalidomide's stereoisomerism was detected or taken seriously in any form at the beginning of its development. The product sold was a racemate.

be resolved empirically due because metabolic activation[14] further complicates a situation that is anything but simple to begin with.

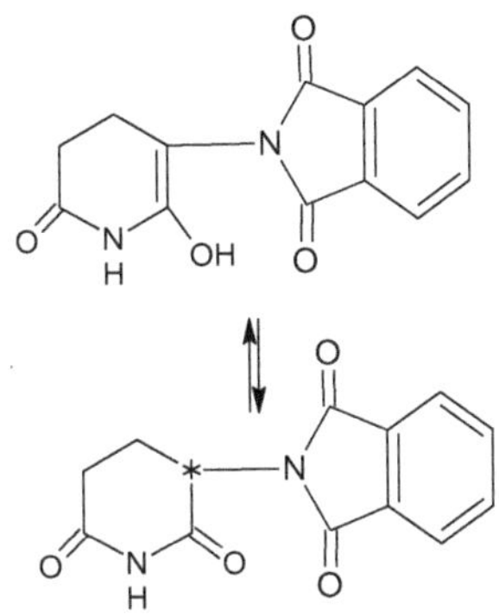

There is a keto form and an enol form of thalidomide that are in equilibrium (keto-enol tautomerism, mentioned further above in the example of acetone) and that, in contrast to the enantiomers, can directly change into one another. Only the keto structure forms stereoisomers (the asymmetrical carbon atom is indicated by an asterisk). As a result of the tautomerism the two forms change into one another under physiological conditions until a racemate is present because the formation of the keto form does not proceed in a stereo-specific manner. Because all known forms of the agent are poorly soluble in water this racemate formation process can take from a few minutes to several hours depending on the composition of the aqueous matrix.[15] Additionally, one must consider that all variants of the active ingredient are subject to hydrolytic separation in the body so that not only its substance identity but also the actual active species remain indeterminate.

In contrast to other cases in scientific research and development in which the goal is "only" to advance knowledge, poor science in pharmacological research is invariably ethically poor science as well because the risks accepted have immediate effects on humans and laboratory animals.[16] The "shivering cage test" is any-

14 In pharmacology and toxicology, "metabolic activation" is understood to mean the biochemical modification of the absorbed substance when it comes into contact with appropriate tissue. Often the active entities are not identical to the preparations administered.

15 Cf. Knoche & Blaschke (1994) and, for the broader chemical context, Roth (2005).

16 Roald Hoffmann, 1981 Nobel Prize winner in chemistry writes: "[...] yes, this is abysmal science. And whereas science as a system for gaining reliable knowledge works in spite of instances of poor-quality experimentation – it will easily survive sloppiness, hype, and even fraud – the kind of science that touches on human lives cannot afford to be bad. The thalidomide disaster should not have been allowed to happen." (Hoffmann 1995, p. 136)

thing but convincing and its results have since failed to be verified in any way. The toxicological data specified in the only original paper of 1956 are without exception questionable. Although until about 1961 probably no one from the group of synthesizers, producers, testers, and distributors can rightly be accused of having knowingly acted in disregard of the drug's embryopathic effect, the purchasers, especially the female purchasers and their children, became "canary birds in the mine of our future."[17] When one considers that in the nearly four years of unrestricted sale of the agent approximately 5 million people consumed a total of 300 million doses, the obvious conclusion is that many of these "canary birds" acted actively and in the absence of a seriously verified indication (the drug was initially available without a prescription). Whereas this in no way indicates consent to all conceivable risks, it does show that a certain lack of concern regarding the use of medications was prevalent in the 1950s. It is also doubtful that this lack of concern really disappeared after the flood of regulation in the wake of the thalidomide catastrophe.[18] In the United States the determination and steadfastness of one individual was able to avert an epidemic catastrophe, although it should be noted that Frances Kelsey was motivated not by specific knowledge, that is genuine foresight, but by an emphatic precautionary principle. Yet unregulated field testing led to a series of cases: Dr. Ray Nulsen of Cincinnati was in attendance at several stillbirths and live births of deformed babies among his patients between January and May 1961 but failed to disclose having administered the sedative and later even vehemently denied it. A further physician distributor who came upon the sedative's embryopathic disposition as a result of his own lack of caution was pediatrician William McBride, who at least admitted his complicity. He made his discovery not based on scientific research but because he unsuspectingly prescribed the medication.[19]

To summarize, we may conclude that at several points the sequence of acts of commission and omission necessary for ethically appropriate behavior was reversed to achieve the exact opposite effect: Where an act of omission would have been correct, the agents exposed others to risks; where they should have intervened, they failed to do so. Thus it is not the substance itself that by its nature is bad or merely dangerous. To this day, thalidomide has been used successfully, albeit not as a seemingly harmless sedative but as a medication against certain types of leprosy. Additional uses do not appear to be excluded. Here, too, the

17 I have taken this image from Murphy & Goldkind (2005).

18 See the relevant discussion in Kessel (2015).

19 Due to his fame, McBride had an exceptionally successful career, but he overextended himself and ended up losing his license to practice medicine. See Stephens & Brynner (2001) for this and other later phases of the thalidomide story outside of Germany.

full story of the specific substance has not yet been told. The teratogenic side effects, whose origin came to light in the late 1950s in spite of extremely irresponsible and defensive company policy, led to significantly improved regulations for the approval of pharmaceuticals throughout the world a few years later. One direct consequence of the thalidomide catastrophe is that many countries now require new drugs to undergo toxicological testing in several species of animals.[20] Another clearly positive social effect, albeit one that came at great cost, is the significantly improved, more enlightened behavior toward persons with disabilities that has been evident in Germany since the 1960s (see the *Aktion Sorgenkind*).

10.3.2 CFCs and Ozone

The use of chlorofluorocarbons or CFCs has had detrimental effects on the ozone system of the middle atmosphere and in turn has led to political developments throughout the world. Here, too, I will provide a brief outline of the chronology of the case:[21]

(1844) Christian Friedrich Schönbein, professor of chemistry in Basel, Switzerland, reported on his electrolytic experiments in creating the oxygen allotrope ozone.[22] (1881) The Irishman Walter Noel Hartley later identified the absorption lines in the ozone spectrum that now bear his name. (1891) The Belgian chemist Frederic Swarts synthesized the first chlorofluorocarbons. (1913) The French scientists Charles Fabry and Henri Buisson detected ozone in the middle atmosphere. (1922) Ozone measurements were begun at Oxford. (1930) The British scientist Sydney Chapman (1888–1970) published his theory of the ozone layer. (Beginning in 1930) CFCs were used in industry as coolants, propellants, or for extinguishing fires, especially in the United States. (1931) The Dobson photometer was constructed to determine ozone content in all layers of the atmosphere. (1957) James Lovelock developed the sensitive electron-capture detector (ECD) for gas chromatography. (1970) The Dutch meteorologist Paul Crutzen postulated the ozone-depleting effect of nitrogen oxides. (1971) Harold Johnston described the negative influence of injections of nitrogen oxides into the lower stratosphere from supersonic aircraft. (1972–1975) The *Climatic Impact Assessment Program* was established and

20 Note that no one can know for certain whether the animal species used in preclinical studies are sufficiently sensitive for detecting all possible risks and side effects – another effect of ontological underdetermination.
21 This overview is based primarily on the report of the World Meteorological Organization, WMO (2018) as well as Christie (2000), Fabian (1987 and 1995), and Schummer (2020).
22 See Schönbein (1844).

implemented in the United States. (1974) The American chemistry professor F. Sherwood Rowland and his young coworker Mario Molina predicted the catalytic breakdown of stratospheric ozone by chlorine from CFCs. (1974) Lovelock measured the vertical distribution of CFCs after taking samples from an aircraft equipped with an ECD. (1975) Spectrometric detection of the fluorine sink hydrogen fluoride in the stratosphere. (1978) The United States prohibited the use of CFCs as propellants in spray cans. (1981) The United Nations Environment Programme demanded international regulations to protect the ozone layer. (1985) Experimental evidence of the Antarctic ozone hole. (1987) The Montreal Protocol for the protection of the ozone layer was ratified by all UN member countries. (1995) Crutzen, Molina, and Rowland were awarded the Nobel Prize in chemistry.

CFCs do not occur in nature. These are a group of chemicals nearly all of which exhibit minimal toxicity to humans and therefore seem practically predestined for their primary uses as solvents, propellants, and coolants. Where closed circuits are not ensured in these applications, these highly volatile substances easily escape into the environment. Although the earlier production quantities of several tens of thousands of tons annually would seem to be a lot, only quite small concentrations were detected in the atmosphere in the 1980s and 1990s. In the middle northern latitudes, these concentrations were typically in the pptv to ppbv range.[23] The comparatively high chemical stability of these substances (for which the low toxicity is an indicator) means that these substances readily pass through the tropopause at an altitude of 10–15 km (a cold trap of down to −80°C) whereas many other substances, especially water, remain below it in the "normal" weather activity of the troposphere. There they usually disappear from the atmospheric chemical processes into so-called sinks or wander cyclically through the environmental compartments. In contrast to the chlorinated compounds to which they are distantly related, the fluorinated substances do not have any appreciable sinks until far into the stratosphere. Only in the higher layers is the radiation sufficiently hard to separate fluorine atoms from the molecules. Whereas the chlorine atoms, which are more easily separable, enter into catalytic ozone breakdown cycles, the fluorine atoms often rapidly bond to hydrogen (forming the extremely stable compound hydrogen fluoride, HF) or to other substrates and therefore play no role in the catalytic process. Above the tropopause, the human influence of artificial CFCs on the natural system is as severe as it is difficult to detect. At least that is the lesson that can be drawn from history. Now after over 40 years of intensive

23 The abbreviations pptv and ppbv mean 1 part volume per trillion and 1 part volume per billion; ppmv means 1 part volume per million. I have taken all atmospheric chemical information from Fabian (1987 and 1995).

international research on the ozone layer, we have acquired satisfactory knowledge of the essential relationships. Several catalytic ozone breakdown mechanisms have been identified, which are referred to by the main catalysts as HO_x, NO_x, and ClO_x cycles. The main catalysts of these cycles became known in the order of mention. Of these cycles, the decisive one in the present context is the ClO_x cycle. Source gases that have passed through the tropopause such as CH_3Cl (natural) and CCl_3F (anthropogenic) are broken down by shortwave light. Upon encountering an ozone molecule, the radical chlorine atoms released by this breakdown process separate an oxygen atom, forming a ClO radical. This radical can in turn react with another oxygen atom and form O_2 and Cl, whereby the circle is closed and the process can begin anew. A single chlorine atom can pass through this cycle several thousand times.

When the ambient conditions are favorable, for example when it is not too cold and the light intensity is sufficiently high, this catalytic process becomes enormously efficient even though the concentrations of the artificial source gases only lie in the pptv range. Chloromethane, CH_3Cl, is a natural source gas that reaches concentrations in the ppmv range in the middle atmosphere. Natural source gases are also found for the other breakdown cycles, for example laughing gas, N_2O, for the NO_x cycles.[24] Thus it becomes clear that the breakdown of ozone and even the ozone hole have a natural basis – as do other major environmental phenomena such as the greenhouse effect and acid rain. Humans contribute to these phenomena but they do not cause them.

The ozone hole above Antarctica, a brief pronounced decrease in the concentration of ozone in September, occurs because in the Antarctic winter a giant stable vortex of air forms that cuts off the inflow of ozone-containing air from the equatorial region where global ozone creation takes place. During the winter, that is until early September, the temperatures in this vortex are so low that many airborne substances condense to form acidic ice crystals. In spring the sun is higher in the sky and the temperature rises. Within a few hours the ice crystals sublimate and under the influence of the stronger sunlight form, among other things, chlorine atoms that take part in the catalytic breakdown process described above. As one chlorine atom can "break down"[25] several thousand ozone molecules and

24 As with other gases relevant to the climate, human activity influences the N_2O balance, for example from the cultivation of rice. The total worldwide annual emission of N_2O is in the range of tens of millions of tons.

25 For lack of a better expression, I use the customary chemist's jargon here. A catalyst does not actually break a substance down but accelerates its breakdown reaction(s). Cf. the discussion in Chapter 6.

the inflow of ozone takes some time, the reservoir of ozone is very rapidly depleted.

Nobody foresaw the ozone hole phenomenon, not even Molina and Rowland. Until far into the 1990s the situation was like this:

> Although good laboratory measurements were available for many reactions, these were applicable to the stratosphere and mesosphere only with rough extrapolations. This is one major reason why calculated ozone reductions due to anthropogenic effects had to be revised so frequently and often drastically. (Fabian 1995, p. 50)

Earlier measurements had already shown systematic fluctuations in the ozone content of the middle atmosphere, for example above Antarctica. The station operated by the British Royal Society in Halley Bay registered measured values beginning in the mid to late 1950s The measurements above Antarctica in the first three years of ozone registration by Halley Bay were a great surprise: A progression similar to that in the northern hemisphere (Spitzbergen) was expected and predicted, only shifted by six months due to the seasons. Yet from 1956 on, a low point in the Antarctic ozone layer below 300 DU (Dobson Units = 3 mm ozone layer thickness at normal pressure and 0°C) was measured in September/October, and a rapid rise to over 400 DU at times in November. Gordon M.B. Dobson (1889–1976), a pioneer of atmospheric ozone chemistry and the creator of the Dobson Unit, commented about 10 years later on the situation at that time as follows:

> The annual variation of ozone in Spitzbergen was fairly well known at that time, so, assuming a six month difference, we knew what to expect. However, when the monthly telegrams from Halley Bay began to arrive and were plotted alongside the Spitzbergen curve, the values for September and October 1956 were about 150 units lower than was expected. We naturally thought that Evans had made some large mistake or that, in spite of checking just before leaving England, the instrument had developed some fault. In November the ozone values suddenly jumped up to those expected from the Spitzbergen results. (Dobson 1968, p. 403)

The confirmation of the initial results in the following years reassured the researchers as to the reliability of the measurements. Although this does not yet amount to a discovery of the ozone hole, the measured data are a clear indication from today's perspective that similar processes had been occurring earlier. Nonetheless, the same questions about the causes, as Rowland posed them a quarter century later, would have been possible and obvious even when the measurements were begun. However, Dobson regarded the situation as a geochemical anomaly, that is as a fluctuation phenomenon.

The measures enacted based on the predictions by Crutzen, Molina, and Rowland, in particular the Montreal Protocol, which was also repeatedly revised, have demonstrably improved the situation as it stands today. Recent prognoses project a "recovery" of the ozone layer in the next few decades, assuming that no unexpected influences occur.[26]

The significance of the prediction in this case of chemical underdetermination is particularly interesting. How can a successful prognosis be developed in a situation for which many different theoretical explanation models come into question? And: What was correctly predicted in this case? In Carl Gustav Hempel's classic philosophy of science, the prediction is structurally identical to the explanation (Hempel 1977, pp. 40–54). Hempel accordingly distinguishes two forms of prediction. He calls the one deductive nomological and the other inductive lawless or inductive statistical. A chemical example of the first version would be the "prediction" of the empirical formula of a saturated hydrocarbon with 4999 carbon atoms: The formula is $C_{4999}H_{10000}$. If such a substance exists, then this statement of its composition will necessarily be correct. The second version is the more problematic one because it is significantly closer to complex scientific practice. Hempel feels that an adequate answer to a why question requiring an explanation is always a possible answer to the corresponding epistemic why question. On the other hand

> a lot of experimental data can strongly *support* the supposition that the electrical resistance of metals increases with temperature or that a certain chemical inhibits the growth of cancer cells without providing an *explanation* for these supposed empirical uniformities. (Hempel 1977, p. 45; original author's emphasis)

These predictive conclusions are more inductive and thus fall under Hempel's second version of prognoses. In her major study of the ozone problem Maureen Christie, like Hempel, differentiates "'prediction' as deduction of the consequences of a theory" and "'prediction' as an attempt to foretell and describe the future behaviour of a system under scientific study" (Christie 2000, p. 73). The author regards both versions or aspects, "entailment" and "prophecy," as present in the statements by Molina & Rowland (1974). A number of important facts were known, albeit from the laboratory perspective, and did not have to be predicted and confirmed. These included such things as knowledge of the lack of a sink for CFCs, their extreme stability in the atmosphere, the splitting off of chlorine atoms under harder ultraviolet radiation, and the breakdown of ozone catalyzed by these chlorine atoms. Little or nothing was known about the actual relationships regarding source gases containing chlorine, and, as was mentioned, the ozone hole

26 See the executive summary of the 58th Ozone Report, WMO (2018, p. 16) and Solomon et al. (2016).

phenomenon really only came to be understood in the mid-1980s although it has a natural basis. Some of the more quantitative predictions of the pioneers of ozone research could not be verified. In other words, they were shown to be false. Chapman's 1930 theory based solely on oxygen predicted significantly more ozone and suffered this fate:

> [...] a philosopher of science cannot regard any theory as correct if it has entailments or consequences that are not borne out by observation. Chapman's theory made a clear prediction of stratospheric ozone levels that were roughly twice the levels that were actually observed. It therefore had clear empirical failings, and in this sense it was 'falsified' or 'wrong'. (Christie 2000, p. 16)

In the case of the Chapman cycle, it was soon recognized that there must be additional ozone breakdown mechanisms. Disproving individual predictions does not necessarily make the entire theory obsolete if multifactorial influences are in play. And multifactorial influences are the rule in chemistry, especially in environmental chemistry. Christie states:

> Prediction (qua prophecy) in this case played a rather different role. It was important that the prediction in this sense was broadly right, but mattered little that it was not right in detail. It was the fact that the particular prophecies drew attention to a possible area of public concern, that provided impetus for both scientists and policymakers to tackle problems in this area. (Christie 2000, p. 91)

Thus if the entailment aspects of an explanation for a chemical problem fail, the prophecy aspects can remain valuable. The author thus arrives at a positive, quasi moral evaluation of the works of the ozone pioneers because the right measures have finally been enacted to "heal" the depletion of ozone. Joachim Schummer sees the matter similarly. He regards Molina & Rowland as exemplary of morally appropriate behavior. He describes the ignorance displayed particularly by the experts in the sciences of substances around 1974 with the following words:

> No chemist would by then imagine that the ozone depletion potential is a property of chemical compounds, which it is of course. (Schummer 2020, p. 45)

Regardless of the fact that Frank Sherwood Rowland was himself a chemist, this evaluation is undoubtedly correct: In chemistry the view is often advocated that the detection of its composition and the identification of its submicroscopic structure identifies a substance for all time.[27] Yet an adequate characterization of a

27 As we have seen, this is unfortunately the case on occasion in those areas of essentialist analytical philosophy that are far removed from empiricism.

chemical entity (Latour's *sturdy envelope*) is not identical to its complete structural description. The encounter with other substances, especially those that are very remote or do not yet exist, can initially escape the focus of chemical research entirely. Schummer characterizes the ideal situation in the framework of the responsibility debate as follows:

> A responsible chemist would take all safety measures, including the employment of safety devices against unknown risks, and undertake research to identify potential threats. Because the active search for potential hazards belongs to the actual habit and capacity of experimental chemists, unlike for instance the theoretical physicists, it would be expected from them also in cases outside the lab. (Schummer 2020, p. 48)

An entire industry of institutions and expert groups of varying sizes have turned their professional attention to these aspects, ranging from the production of laboratory equipment to education and advanced training in laboratory safety to insurance companies and law practices. Conflicts occur where the classic manufacturing program – the Berthelot principle – encounters possible risks. Schummer states:

> The traditional focus in chemistry has been on studying how to make things in the laboratory. In contrast, researching possible hazards requires the study of open systems that usually transcends disciplinary boundaries and is thus intellectually more demanding. Although a growing number of chemists have engaged in that kind of research since the 1960s, they are still a minor part, and marginal in the dominant self-image of chemistry, despite the extraordinary societal impact of works such as that by Molina and Rowland. (Schummer 2020, p. 51)

In the setting of the discussion examined in this chapter, a discussion otherwise characterized by widespread approval, the question arises as to which open systems should be examined. No one could have imagined which processes would actually take place in the ozone layer and, again, that something like an ozone hole exists. Crutzen, Molina, and Rowland have opened the door, and the rooms behind will have to be studied for a long time although we can already discern the floor plan.

Situations similar to those examined in the previous sections can be found based on additional cases. DDT, the phthalates used as plastic softeners, and the organic tin agents in antifouling marine paint have also given rise to unpleasant surprises that fit into the framework of what I would describe as ontological underdetermination. A situation, in a certain sense inverse, that also belongs in the relationship mentioned above occurs due to the plastic refuse in the environment. Microplastic and plastic packaging materials are incorporated by animals where they cause harm despite being "harmless" and chemically neutral in the conventional sense for the very reason that they cannot be metabolized.

10.4 Conclusions

Not only theoretical interpretations are subject to historicity but the respective stock of epistemic objects in the various chemical spaces as well. Recently, the expression "chemical space" has become a terminus technicus in the history and philosophy of chemistry, cf. Restrepo (2022). Therefore, in addition to epistemic underdetermination which normally leads to pluralism, there exists another version typical of chemistry: ontological underdetermination.

Who can predict the risks and side effects that can arise from this ontological underdetermination? There is only one genuinely honest answer to this question: Nobody. Whether it be local, global, or both, the epistemic frame of reference that might be relevant is initially unknown: No one could foresee that there would ever be a solvent for gold; no one could foresee thalidomide's terrible embryotoxicity; no one could have known that CFCs, harmless in the laboratory, would have ecotoxic side effects when released from their closed circuits. This negative perspective is in no way intended to belittle the significance of environmental chemical studies: The evaluation of ecologically relevant characteristics is usually performed on individual substances, yet their effect is invariably relational. What relation is present and how many digits it has is initially not at all clear with problems of environmental chemistry.[28]

It is obvious that substances that are in some way active very probably have additional effects, that is to say, side effects. No one should continue to dream the dream of a drug with exclusively positive effects, just as there are no exclusively "bad" substances. Paracelsus with his dosage thesis was correct in both regards. As varied as identified hazardous substances[29] are (like the thalidomide and CFCs discussed here) one possible common characteristic of those who manufacture, examine, use, or distribute them can be fateful: carelessness. If we handle substances in our environment carelessly, the risk increases; this applies especially, but not solely, to chemists. The increasing risk may not be limited to this group of persons itself, as in laboratory work, but may affect others as well; unfortunately this increases the chance that this carelessness can turn into indifference, as the thalidomide catastrophe has shown us.

Obviously, underdetermination in chemistry is closely related to emergence, and particularly so if an ontological change is involved. Lee McIntyre (2007) argues for an epistemological interpretation of reduction and emergence. We have seen,

---

28 See Korte (1987), Friege et al. (1998), and Scheringer (1999) for additional known cases and possible examination methods.

29 Cf. the accounts of hazardous substances by Jens Soentgen (2019).

however, that at least the latter has some inevitable ontological connotation to it. In some cases, a posteriori explanations seem possible in general. The uncertainty of predictions remains, however.

The ontological underdetermination in chemistry has no simple and predictable solutions. In any case, in addition to professional expertise on the part of all involved, attentiveness, care, and a sense of responsibility are required to reduce risks and conflicts in handling substances. Surprises can never be excluded; yet that includes positive ones.

Fig. 26: Theory or Practice – Reduction or Pluralism?

11 Synopsis – Fundamentals of Metachemistry

In his monograph on the philosophy of science entitled *La philosophie du non*, a work emphasizing real-world practical experience, Gaston Bachelard introduces the concept of "metachemistry."[1] He feels that this concept is justified and that it relates to metaphysics in the same way that chemistry relates to physics. He continues as follows:

> Metaphysics could only have one concept of substance because the original understanding of physical phenomena was content to examine a geometrical solid body characterized by general properties. Metachemistry must benefit from chemistry's knowledge of the various forms of behavior of substances. It must benefit from the fact that actual chemical substances are more products of engineering than bodies encountered in reality. (Bachelard 1940, p. 68)

Thus chemistry strives primarily to achieve an understanding of actual substances, which is why there is fundamentally limited scope for conceptual abstractions of the sort customary in physics. As Joachim Schummer has found, what predominates in chemistry is an experimentalist entity realism. In the concluding chapter of this study, I now try to pull together the essential aspects of metachemistry examined here.

(1) Chemistry is the study of the substances. It is inexpedient to attempt to define it with supposedly greater precision as the science of molecules (which is often done, especially in the philosophy of science). Nor is it accurately characterized as the study of the combinations of (modern) elements. The claim that it is the scientific discipline that concerns itself with the causes and effects of the shedding, acquisition, or distribution of electrons between atoms or molecules and with the relationships between the energy levels of such electrons within the atoms or molecules is completely unsuitable as a general definition. All the attempts to import elements of the latest theoretical lines of thinking to arrive at a suitable characterization may have some justification as component parts of an explanation. Nonetheless, they fall short when they negate the stock of empirical experience. The latter is the starting point and the basis of all theoretical efforts.[2] In any case,

---

1 For a discussion of this concept cf. Nordmann (2006), Bensaude-Vincent & Simon (2008), and Bensaude-Vincent (2012). Also pertinent is Frevert (2014).

2 Jaap van Brakel rightly emphasizes this long-recognized positivist principle in particular for the realm of knowledge of the substances. He explains this very vividly: "If quantum mechanics turns out to be wrong, it would not affect *all* chemical knowledge. If molecular chemistry (chemical atomism) were wrong, it wouldn't disqualify *all* knowledge about, say, water. What there is, are

https://doi.org/10.1515/9783111401041-014

chemistry is not a part of the program of certain fields of theoretical physics that seek to identify the ultimate building blocks and forces of the world. It concerns itself not with matter but with substances. These exhibit spatial dimensions, and in most cases a readily measurable mass. Beyond this, they reveal their nature through a variety of relational associations (characteristics, dispositions) among each other. Substances often manifest themselves to us via their thing-like character, yet they themselves are not necessarily things.

(2) The emergence of that which is new is – to a significantly greater extent than in any other scientific discipline – a central aspect of chemistry. Material novelty can arise from intentional production by humans, yet it can also occur completely unintentionally when substance entities come into contact in uncontrolled or uncontrollable situations. Underdetermination in chemistry exhibits not only the conventional epistemological characteristics but also ontological characteristics. This is an essential and creative feature of chemistry: Other scientific undertakings can influence their range of objects only to a far lesser extent. And new objects or phenomena repeatedly bring with them anew the classic problem area of explaining and predicting their characteristics. Can we reliably predict and thus explain the result of the chemical reaction of hydrogen with oxygen from everything that we know about these two elemental substances? Can we identify the characteristics of water from the knowledge of its constituents? The clear answer is no. Except for the mass, the chemical characteristics of the product in the strict sense are *emergent:*[3] Only the mass of the product, which is nearly entirely determined by the atomic nuclei, behaves approximately additively and is therefore predictable.

In both science and *Lebenswelt* we have been accustomed to usually uncritical references to submicroscopic entities and mechanisms, and thus to constantly switching conceptually between the manifest macroscopic world and the world of the non-observables. Yet this aspect remains indisputable: The starting point of all chemical descriptions and interpretations is the empiricism of substance. And this empiricism of substance abruptly follows and is intimately associated with perceptible sensory experiences, although it is in no way limited to them. Even philosophers of science who usually advocate certain forms of microstructuralism admit the usefulness of emergentist thoughts (cf. Hendry 2006, 2017, 2021).

---

chemical and physical descriptions of macroscopic entities, whose identity conditions are grounded in the end, in the manifest image." (van Brakel 1997, p. 273; original author's emphasis)

3 I introduced the notion of "emergence" in Chapter 1. It seems that "downward causation" or "strong emergentism" are in place when it comes to the discussion of the relation between composition and properties of chemical substances. I am confident to have clarified in the present study that the hope to show the opposite – some call it strong reductionism – is indeed futile.

(3) Although trivial when considered dispassionately, the following fact appears to me to be enormously important in the present context: There is a chemistry without atoms as the studies by František Wald and Wilhelm Ostwald have clearly demonstrated. Only a few modern textbooks of chemistry examine the concept of substance in greater depth, and since Ostwald's *Prinzipien der Chemie* not a single one in the whole world has based its fundamental conception or exposition on this concept. In the vast majority of cases, the differences between mixtures and pure substances (elements and compounds) are mentioned merely briefly, usually followed by an introduction to atomic physics.[4] This automatically conveys the impression that the world consists of atoms and that the normal path of acquiring knowledge in chemistry is thus from "bottom to top." The flood of television images to which people today are inescapably subjected from a young age contributes to reinforcing this naively compositional concept (cf. Figures 3 and 18).

In reality, that is in chemical practice, the acquisition of knowledge proceeds in precisely the opposite direction. Hardly anyone works with individual particles, and when they do – as in the case of experiments with super heavy elements (SHE) – this work is only possible due to the extremely highly developed design of the experimental conditions, which as well draws on (at least implicit) knowledge of substances. The particle-centric image of reality that we call atomism is a conceptual construct and not something that one works with *in practice.* In many applications of chemistry, if not in its core fields of endeavor, there is simply no need for particle-centric theories. The productive success of a synthesis is indisputably the focus of interest. Chemical practice involves specimens of substances; reflection on the relationships employs simplified abstract images of these specimens. Many of the fundamental achievements of chemistry and its precursors, such as the methodology of substance preparation, substance characterization, hylotropism (albeit applied only implicitly), and even the periodicity[5] of element characteristics, have been attained entirely without atomist theory. An interesting point in this regard is the circumstance observed by Friedrich Paneth that even the most physicalist def-

---

4 Even the standard university textbook of many earlier generations of students in the German-speaking countries, the otherwise exceedingly comprehensive Holleman-Wiberg (1964), which devotes 10 pages to the concept of substance, gives the chapter the title Atom and Molecule and in so doing follows the main thrust of atomism.

5 I have allowed myself the luxury of largely ignoring the periodic system of the chemical elements in this study of the philosophy of chemistry. This is not due to any fundamental neglect much less disdain on my part but to the chosen substance perspective, which in terms of methods and systematics comes before the examination of current aspects. I refer the reader to the works of Eric Scerri, in particular Scerri (2007), for an introduction to this field.

inition of an element does not attempt a complete characterization of the atom with the atomic number but only of part of the atomic nuclei, namely the protons.

(4) Professionals in chemistry "make" their epistemic objects themselves. That is to say, they seek to find specimens of suitable source substances from things in the environment, clean these specimens, subject them to certain manipulations and alterations, and evaluate the success of their methods by comparing the specimens to known substances (or their characteristics). Trust in a result thus obtained increases with the number of verified characteristics of these control substances. It is typical of chemical work that each successful analytical procedure involves synthetic aspects.

The manufacturing character of chemistry is not limited to substance entities; species that are not produced in the strict sense and can be poured into containers are included as well. Even SHE entities can be manufactured artistically and systematically, although they do not exhibit the full character of substances and belong more to the fringes of chemistry.[6]

To what extent do the insights obtained in the laboratory, that is to say "artificially," have to do with those obtained in the environment (in "nature")? The notion of natural kinds is ambivalent. Essentialism proves inadequate upon closer examination of the actual practice of chemistry; it is at best merely a particular form or border case of descriptivism. Laboratory experts can precisely define the concept of *purity* experimentally and quantify characteristics that relate to non-observables. I find it hard to understand why some philosophers then choose precisely the designation "natural" (in the sense of "independent of humans") to describe the results of these highly creative and artificial activities, such as the measuring of nuclear charges. On the contrary, I would like to emphasize that we can acquire knowledge about the materiality in nature only via interventions in substance systems, that is via the practiced systematic art of chemistry. A certain fundamental artificiality or, perhaps better, a certain artistic skill, is thus an essential feature of the science of substances.[7]

---

6 As an example I refer the reader to the fascinating work of a large international research group who has presented measurements of 7 (!) atoms of the element hassium, which the authors without any further explanation regard as the *chemical* characterization of this heavy element (Düllmann et al. 2002).

7 Peter Janich goes as far as to state: "Nature can only be understood as a technical model" (Janich 2006, p. 177). This correlates with the interpretation by Böhme & van den Daele (1977, p. 188) quoted above that the difference between natural things and technical artifacts disappears.

(5) Chemistry is pluralistic. In theory and practice, there is more than one way to succeed in chemistry.[8] Success here means first and foremost the realization of chemical phenomena and the manufacture of products. Scientific pluralism can be treated in different conceptual ways: One can differentiate epistemological and ontological variants, theoretical formal and practical methodological forms, and also descriptive and normative approaches. Because of the peculiarities of chemistry that have been examined in this study, it is best to place the focal point of the discussion of pluralism more strongly on the reality of substance. "Substances occur," states Jens Soentgen, and they are the main focus of chemistry. Naturally, this does not mean that the science of substances is completely free of theory. However, good science and of course the history and philosophy of science as well cannot be constructed solely by proceeding from the supposed endpoint if the goal is to achieve a comprehensive and accurate understanding of the actual science of chemistry. Current theories about elementary particles and electromagnetic fields may have their occasional benefits for chemistry, yet they tend to lead one away from its "middle" reality.

Jaap van Brakel has demonstrated one way of illustrating pluralism in chemistry. The following programmatic statement appears in a paper on the neglect of the philosophy of chemistry:

> [Al]though there is now extensive literature in the philosophy of science with references to detailed chemical examples, the number of studies specifically addressing the question of what makes chemistry special or, respectively, what constitutes its autonomy, has remained very small. My own philosophical interest lies in defending antireductionist forms of a pluralism and it is based on the fundamental priority of our manifest ideas of the world (such as Husserl's "Lebenswelt," Heidegger's "Dasein," and Wittgenstein's "form of life") over any scientific image. (van Brakel 1996, p. 22)

If there is any asymmetry in the system of scientific knowledge, then it is the priority of the "manifest world" concerning the starting point of every individual examination. Although it is essentially trivial, it must repeatedly be emphasized to self-proclaimed "realists": The current scientific image is always downstream and it can change. Van Brakel is very critical of philosophical interpretations emphasizing relations between discussions such as reduction, supervenience, and emergence. For example, the scientific description of modern chemical elements

8 Cf. our Ruthenberg & Mets (2020). I leave the question unanswered as to whether there is an ontological pluralism. In the *one world one science* debate it is often unclear whether the focus is on epistemological or ontological relationships. The physicalists' claim that everything is ultimately reducible to the matter described by physics, whatever that may be, is on the one hand cheap and on the other hand philosophically questionable if not dishonest.

begins with the observation, preparation, and examination of specimens of substances. The knowledge that arises in this manner is multifaceted and is essentially acquired in an "intercultural" context: What begins with prospecting for gold, continues with metallurgic purification, proceeds on to a variety of chemical experiments, and does not yet end with the measurement of so-called physical values, is based on the pluralistic gathering of fragments of knowledge. All these fragments have their separate specific meaning. From an ontological perspective one could assign priority to the atomic number, whereas from an epistemological perspective, and that is the experimental one, the focus is on the substance. Jaap van Brakel rightly emphasizes the equal footing of the different views – such as the macroscopic, microscopic, and submicroscopic ones – and rightfully rejects the colonization of certain forms by supposedly more fundamental forms.

Joachim Schummer's justified criticism of "substance-less" philosophy may be found in his discussion of methodological pluralism in chemistry (Schummer 2015). In it he finds that the fragmentation of the modern natural sciences in many component disciplines is itself an indication that modern sciences fundamentally tend toward pluralism. He gives two specific reasons why it is unavoidable for chemistry. First, the tasks of chemistry are many and varied, for example in the classification of substances employing its qualitative concepts, the explanation and prediction of the behavior of substances utilizing its quantitative approaches, and the operational concepts of chemical synthesis. For this reason, it is difficult to identify a higher-level goal such as "scientific truth" in chemistry. The second specific reason for chemical pluralism according to Schummer is found in the unbridgeable gap between the concepts employed in chemistry and their epistemic objects. In contrast to broad areas of theoretical physics, the description of substances sets narrow limits to abstractions:

> […] chemistry has addressed the realism/nominalism issue, which is sometimes misleadingly called the issue of natural kinds, by experimental means, largely unnoticed by mainstream philosophy. Rather than only adjusting our concepts by creating experimental systems to the world as it is, chemistry also adjusts the world to its concepts by creating experimental systems that best fit its frameworks. (Schummer 2015, p. 62)

By virtue of its manufacturing character chemistry changes the world, on a small scale or large scale, locally or globally. Insofar as it applies to both the scientific and philosophical study of substances, pluralism is unavoidable from a methodological perspective.

Hasok Chang goes one step further toward a normative interpretation of general scientific pluralism:[9]

> [M]y position is that if we should find a field of science which is quite monistic, then that is quite likely not healthy, and we should consider reforming it. (Chang 2012, p. 269)

He describes his approach as "active" pluralism (he uses the term *active normative epistemic pluralism*). In contrast to the conventional philosophies of science and their search for ultimate universal laws, this active pluralism is intended to cultivate the contest of competing systems of practice; Chang expressly avoids using Kuhn's related paradigm concept. Significantly, Chang draws his case studies particularly from the substance-related sciences. We have already encountered his interpretation of the atomism debate in the 19th century in connection with the discussion of underdetermination. Active pluralism amounts to identifying and emphasizing the advantages of tolerating different scientific descriptions and the advantages of mutual interactions among competing systems of practice. Here, variety takes precedence over uniformity and coexistence over reduction. Chang even goes as far as to describe the concept of phlogiston, which in the conventional history of chemistry is portrayed as incorrect, obsolete, or not borne out, as worthy of retention.[10]

A multitude of other examples can be cited to illustrate and support the notion of a fundamental pluralism in chemistry, almost regardless of where one looks. I have discussed the element concept in greater detail: The "classic" characterization according to Lavoisier coexists with the "physicalist" description using the atomic numbers. Whether or not one accepts Paneth's reduplication of the element concept ("simple substances" and "basic substances"), it is nonetheless clear that there are many ways to characterize chemical elements. Paneth himself reads Bachelard's *Le Pluralisme cohérent de la Chimie moderne* and immediately after its publication in 1932 reacts as follows to how it represents the situation in chemistry as "oscillation between pluralism in one part and reduction of the plurality in the other part":

---

9 Paul Feyerabend is an early supporter of scientific pluralism in *Against Method* (1975). Although by his own admission a supporter of pluralism in chemistry himself, Estonian Rein Vihalemm, one of the pioneers of the modern philosophy of chemistry, criticizes Chang's normative approach; see Vihalemm (2016) and confer the important recent paper collection by Mets et al. (2024).

10 "When Odling [1829–1921] and others saw phlogiston as the predecessor of chemical potential energy, and when Lewis [1875–1946] saw phlogiston as the predecessor of electrons, what they were doing may have been whiggish but it was certainly not pointless. And their insights provide a sufficient answer as to why I am not going to try to bring phlogiston back to modern chemistry – it is already here!" (Chang 2012, p. 65)

> Rather than calling it oscillation, it seems to me that it would be better to speak of simultaneous development in different branches. The reduction to unity (or quadruplicity)[11] has been successfully achieved by physics; chemistry, however, will probably preserve [sic], as long as there is a science of chemistry in the present meaning of the word [...]. (Paneth 1936, p. 70)

The obvious impermanence of the reductionist physicalist representation (not only from a present-day perspective) did not prevent the experts at the time from seeking the primary significance of scientific interpretation precisely there, and many contemporary experts in the sciences and philosophy have continued to do exactly that to this day. Even Paneth himself, despite his many years of advocating the autonomy of the science of substances (see Paneth 1916), places the higher value of scientific interpretation in the realm that he had once termed "theoretical." The perspectivism he suggested is thus not one of equal footing, and chemistry can persist only as long as this prioritized theoretical realm allows it. I feel such a pluralism is lopsided because it disregards actual scientific practice. As I, following Jaap van Brakel, have attempted to show, we must seek the basis for knowledge about chemical entities, and thus knowledge about chemical elements as well, in the manifest realm of the real world.

One can also observe similar developments with other important chemical concepts. The previously mentioned concept of electronegativity is one where repeated attempts have been made to conceptually place it more firmly in the microphysical realm. Linus Pauling established modern electronegativity by astutely comparing thermochemical data (dissociation energies of covalent compounds). It was found that a virtual "partial ionic energy," the "ionic character of the compound," can be calculated from the square of the differences in the electronegativities of the types of atoms present in the compound. When the values for certain elements are standardized (2.2 for hydrogen, 4.0 for fluorine), values between 0 and about 3.3 (see discussion above) result for the differences in the electronegativities. Subsequent attempts to establish an electronegativity scale made use of values such as dipole moments, ionization energies, and electron affinities as well as spectroscopic data. Interestingly, the core Pauling scale is still used, with additions for elements not included by Pauling as well as a few numerical improvements. These attempts are evidence of a trend toward greater abstraction that, as we have seen, is typical of chemical concepts: The more "substance-related" the approach, the more complicated the path to the electronegativity scale; and the more microphysical and quantum-theoretical the method, the greater the concep-

11 Contemporary particle physics distinguished four things: the proton, two electrons (±), and the neutrino.

tual distance to the substance characteristics becomes. Thus there are approaches that at certain points lead to counterintuitive results: According to the atomic physics formula developed by Robert Mulliken[12] one can even assign electronegativities to the inert gases, which is essentially absurd. And yet following Hasok Chang one can regard even these approaches as valuable for further scientific discussion. As there is no single productive path to describing the mereological projection of the "chemical" behavior of atoms concerning electrons, cultivating a variety of approaches from different systems of practice is expedient and advisable. Electronegativity, a chemical concept that is not precisely defined yet nonetheless quantifiable, has been enormously successful precisely because of its theoretical elasticity.

Additional evidence of the pluralistic character of chemistry may be found in the area of acidity: *Protonism* (Brønsted[13]) and *electronism* (Lewis/Pearson) continue to coexist without anybody racking their brains as to why the same words "acid" and "base" are used in both systems of practice. The same pair of notions has measurable substance correlates in one system, whereas in the other it refers to theoretical entities.[14] The coexistence of the two concepts contributes to the variety of possibilities of interpretation in chemistry, and this fact represents a valuable commodity.

(6) Plato and Kant are probably right in maintaining that humans will never be able to grasp the ultimate nature of the world. However, we do not have to understand Kant's "transcendental difference" in a strictly categorical sense (as Friedrich Paneth did). I prefer to interpret it as a kind of heuristic concept, in that it leaves open that there might be some scientific progress. Kant himself experienced such progress in chemistry (see Lequan 2000 and, particularly, van Brakel 2006). At the beginning of this study and in Chapter 9 I described Kant's view of contemporary chemistry. Georg Ernst Stahl, whose work Kant repeatedly refers to especially in his *Critique of Pure Reason*, defines chemistry as the art of separa-

12 According to Mulliken (1934), the electronegativity of an element A results from the mean value of its ionization energy I_A and electron affinity A_A: $\chi_A = \frac{1}{2}(I_A + A_A)$. He also calls this *electroaffinity* (cf. Abegg's publications). The astonishing simplicity of this approach is matched by a high degree of mathematization divorced from empirical. See Ruthenberg & Martínez González (2017) for a critical discussion regarding this point.

13 In a certain sense, Thomas Martin Lowry can be called a protonist, too, but he hasn't been a co-creator of the respective theory, as the customary Anglo-american narrative has it, cf. Ruthenberg (2023).

14 Lewis acids and bases are not measurable as such.

tion.[15] Toward the end of that *Critique*, Kant shows a general parallel to the analytic-synthetic ideal of the science of substances evident in Stahl's writings:

> What the chemist does in the analysis of substances [beim Scheiden der Materien], what the mathematician[s] in pure mathematics [ihrer reinen Größenlehre], is, in a still higher degree, the duty of the philosopher, that the value of each different kind of cognition, and the part it takes in the operation of the mind, may be clearly defined. (Kant 1781, p. 472 or B 870)

What Kant then describes as an "extremely significant" isolation of the constituents – material in chemistry, conceptual in philosophy – has assumed the proportions of doctrines in both disciplines.[16] The formation of the modern notion of an element occurred in the 18th century and culminated in Lavoisier's work. Analysis, in the sense of separation of a substance, increasingly moved conceptually and operationally into the focus of chemical interest. Other important concepts of chemical examinations such as chemical affinity fell behind compositionism for about a century.[17] About ten years after the publication of his *Critique of Pure Reason*, Kant verifiably reacts to developments in chemistry: In the preface to the *Metaphysics of Morals* of 1797, he claims that the "chemist pursues only one chemistry," namely that of Lavoisier (whose *Traité Élémentaire de Chimie* had been published in 1789).[18] By this he meant the new element doctrine which generally was combined with the notion of separation, "isolation," and still is to this day.[19] As was mentioned previously, Kant's 1786 evaluation of chemistry as an "improper" science does not arise from any lack of respect for chemistry.[20] When he refers to chemis-

15 Cf. the treatment of self-descriptions in the introduction to this study and Stahl's definition cited in Chapter 5.

16 By this I of course mean analytical chemistry and the primarily Anglo-American analytic philosophy. Note that I have argued here for the claim that at least the former cannot do without synthetical aspects.

17 The study of affinity (today one would perhaps call this field reactivity) blossomed anew with the works of Guldberg & Waage and Wilhelm Ostwald, who presented attempts to quantify affinity.

18 In *Prolegomena to Any Future Metaphysics* of 1783, his elucidation of the *Critique of Pure Reason*, Kant says that the *Critique* is to the customary school metaphysics as chemistry is to alchemy or astronomy is to prophesying astrology. (cf. Kant 1783, p. 125/Akademieedition: p. 190) Although he classifies alchemy together with astrology rather disparagingly as pseudosciences, it is obvious that he has followed the contemporary science of substances with interest and goodwill.

19 Lavoisier's definition of an element (occasionally the "Boyle-Lavoisier" definition) normally goes as follows: An element is a substance that cannot be subdivided into other substances. No mention is usually made of the fact that fulfilling this operational regulation requires detection of the resulting product or products and thus synthetic and preparatory steps.

20 "Any whole of cognition that is systematic can, for this reason, already be called *science*, and, if the connection of cognition in this system is an interconnection of grounds and consequences,

try as a systematic art, Kant primarily means its empirical character which only allows partial application of formal mathematical-logical methods of explanation. Empirically acquired knowledge cannot achieve the status of *proper* scientific character from Kant's perspective. Yet if not by empirical means, then how else is one supposed to arrive at the limits that, even for Kant, nature necessarily entails?

Hans Cornelius (1863–1947), with a doctorate in chemistry and a university lecturing qualification (Habilitation) in philosophy,[21] rejects Kant's "downgrading of empiricism." His fundamental attitude is that of an empiricist.[22] He feels that every scientific activity leads to "the epistemological line of inquiry" (Kant's "What can we know?"), but:

> Even the epistemological examination cannot lead to a conclusive result as long as the naturalistic premise of the *"thing-in-itself as the cause of the phenomena"* is openly or implicitly assumed. (Cornelius 1921, p. 166; original author's emphasis)

Whereas Cornelius does not develop any specific philosophy of chemistry, the selection and treatment of his examples, often drawn from chemistry, are nonetheless very illuminating. In his critique of Kant, as in the quotation above, he focuses primarily on the "thing-in-itself." He feels this concept is a dogmatic dictum and emphasizes the danger that it may give rise to a dogmatic idealism for which the epistemic objects only exist "for us." Among other things, this would turn concrete substances into abstract entities (a view which, as we have seen, Friedrich Paneth advocates).[23] In contrast, Cornelius feels it is self-evident that stable knowl-

even *rational* science." (Kant 1786/2004, p. 4; Kant 1786/1982, A V; original author's emphases) Cf. Höffe (2011), Carrier (1990).

21 Cornelius was a full professor (Ordinarius) of philosophy at the University of Frankfurt (1910–1928), where his students included Max Horkheimer and Theodor Adorno.

22 By "empiricist" I do not mean "logical empiricist," but "practical empiricist." Cornelius makes no reference to the philosophy of the Vienna Circle, although he does to the work of Ernst Mach. Note that even František Wald mentions him (probably without having read anything by that chemist-philosopher).

23 In his work *Transcendentale Systematik*, the preface begins as follows: "There is probably no error in the theoretical realm that has today enjoyed greater status and wider distribution than the opinion that our knowing is inflexibly denied access to the "realm of that which is being and eternally true." Given that the line of inquiry of practical philosophy and its solution in *Kant* depends on this error, it can be seen that it has more than merely theoretical significance." (Cornelius 1916, p. III; original author's emphasis)

edge exists in the experimental sciences and cites the melting point of a newly synthesized substance as an example.[24] Similarly, he states elsewhere:

> When the chemist has found a new acid and purified it according to certain methods, then with a clear conscience he may record the statement that it, let's say, melts at 40 degrees as a generally valid statement even if he has only made one determination of the melting point. (Cornelius 1931, pp. 210–211)

Aside from the fact that multiple determinations are recommended in chemical laboratories as a matter of course, this statement, especially its epistemological core, is undoubtedly correct. Since the introduction of reliable thermometers in the 19th century, knowledge of the melting point of pure chemical substances developed rapidly, and these data soon entered compilations of analytical tables as substance constants (incorrectly: *physical constants*).[25] Cornelius cites additional cases from chemical practice to distinguish the "correct" induction from the "false" induction on the one hand, and on the other to emphasize the higher value of experiential knowledge compared with "worthless analytical truisms." He mentions the solubility of table salt in water as another case of an empirically determined statement conforming to scientific laws, that is a generally valid statement: "[The statement] that table salt is soluble in water is globally valid for the reason that the name table salt has been determined in such a way that it does not refer to an insoluble body but only a soluble one." (Cornelius 1931, p. 211) According to Kant, analytical statements can never be false, "but analytical statements derived from mere definitions are not scientifically valuable truths but only truisms" (Cornelius 1931, p. 211). The difference between empirically acquired knowledge and "worthless analytical truisms" consists precisely in the fact "that in

---

24 In Section 17 (*Die Voraussetzung der Minderwertigkeit der empirischen Erkenntnis*) of *Trancendentale Systematik* he states: "When the chemist in the laboratory manufactures a new pure substance and has determined its melting point, then he is in any case convinced to have acquired by this determination not merely a simple fact but knowledge valid for all the future that will accordingly be accepted into the archive of science; and, if he has worked precisely, he may be certain of the fact that no future experience will prove his formulation false. The question as to the right of this conviction, that is the question of to what extent his formulation is attributable to material or formal factors, cannot in any case be decided a priori by the claim that empirical knowledge has only relative validity and that accordingly 'chemistry cannot be called a science at all.'" (Cornelius 1916, pp. 30–31)

25 There are hardly any relevant studies on the history of the measurement of melting points, cf. Chang (2004). The practical importance of measurements of melting points has decreased as a result of the great success of spectrometric measuring methods.

our examples *the definition is derived on the basis of experience*" (Cornelius 1931, p. 211; original author's emphasis).[26]

For Cornelius, the division of the world into a world of phenomena and a world of laws is not adequate for dealing with real sciences like chemistry.[27] He describes chemical relationships in a variation of what today is known as scientific realism that one could term non-naive.[28] The technical manipulations of perceptible specimens of substances required for analytical and synthetic purposes counteract naivety. Cornelius describes the differentiation of the realm of the phenomena and the realm of the laws as follows:

> The concept of the persistent object and the concept of its persistent characteristics is to that extent a νοούμενον as opposed to the φαινόμενα, the changing appearances of our senses. *Yet of these appearances that νοούμενον is not more than something foreign separated by an unbridgeable abyss, rather it is nothing other than the order of these appearances determined by our thinking:* the appearances are the individual cases of the general rule given in the νοούμενον. (Cornelius 1921, p. 271; original author's emphasis)

To that extent, the concept of a substance that had once been examined can exist without an actual specimen of it being available. We speak of its existence in the stricter sense when we are observing it or its characteristics at the moment. According to Cornelius, in actual science it is not necessary, even preposterous, to conceive the "substance-in-itself." In other words: Chemistry does not seek analytical truths in the "noumenon." It cannot even do this at all because this "noumenon" is a mental (or conceptual) space whereas chemistry has to do with tangible objects. This does not exclude the possibility of acquiring further experience with a known substance. For example, gold was the same substance before and after the

26 Cornelius continues: "What is valuable are not the analytical statements in which one merely repeats those definitions more or less at length but precisely the definitions themselves and their origin from experience. That there is such a thing in the world as that which specifies the concepts of gold or paper, table salt and 1,3,5-dioxyphenyl acetic acid [sic], and all the legions of other defined concepts of chemical substances: That is the synthetic judgment which is expressed in general in every one of these inductions and which constitutes the scientific value of the inductively derived knowledge. *Induction is nothing other than the formulation of experiences in hard and fast concepts and the general existential judgments expressed with the aid of these concepts.*" (Cornelius 1931, p. 212; original author's emphasis) "Correct" inductions for Cornelius are for example the substance characterizations he mentions; the "false" inductions include the often-cited conclusion that the statement "all swans are white" is generally valid based on observation of a few white swans.
27 The "transcendental difference" is treated in Kant studies in at least two meanings: As a dual-aspect theory and, as attacked by Cornelius, a two-world view or dualism, see Höffe (2011, p. 48).
28 Cf. the experimentalist realism in chemistry mentioned previously and described by Schummer (1994 and elsewhere). Hasok Chang's (2022) novel active realism is very close to this.

introduction of aqua regia, only that it proved to be soluble in this acid. The subsequent determination of the atomic number 79 for gold was also an empirical act, although one might regard this knowledge as ontologically more significant than the "chemical" properties, as the essentialists do. The search for qualitatively and quantitatively better knowledge as science advances seems to echo the notion of the actual essence, the "in-itself," of the specific epistemic objects.

Hans Cornelius, like Paneth a synthetic organic chemist by training, attaches significantly greater importance to chemical practice and thus argues more plausibly than Paneth.[29] As regards atomism he is distinctly more reserved. He does indeed recognize the "transcendental" or "theoretical" character of certain entities and describes the "actual" things as "laws for our perceptions." These perceptions exist in relationships that we come to know through scientific progress. What is important to him is this:

> Yet these relationships can never cease to encompass as well the first sensory perceptions from which alone we have drawn the knowledge of the things. Even the hypothetical constructs that seem furthest from the senses such as atomic nuclei and electrons would lose any meaning for our knowledge of the world without that relationship; they would cease to be theories for *reality.* (Cornelius 1931, p. 216; original author's emphasis)

Thus in Cornelius' thinking, there can be no talk of an "evaporation" of the substance concept as Paneth maintains.[30] On the contrary: This substance concept defines the metachemical and practical starting point of all research and remains the central point of reference in all attainable results. Yet how far we can advance conceptually into the realm of theoretical entities without losing contact with the "sensory perceptions" remains an open question. For the more artificial an experimental design, the more this relationship can be obscured.

(7) So, is chemistry an art? It seems that after the *Metaphysical Foundations of Natural Science,* Kant did not again take up or comment upon his characterization of

29 Cornelius states clearly what essentialists like Paneth often ignore and in so doing often turn the actual practice in the sciences upside down: "We can only study physical and chemical regularities of things, when we *have* the things, i. e., when we already possess the knowledge of certain kinds of persistent laws for our perceptions based on pre-scientific experience." (Cornelius 1931, p. 201; original author's emphasis) That, however, to "have" and to "possess knowledge" of "things," is very often connected to very difficult procedures within the systematic art chemistry, was well known even to an essentialist like Paneth.

30 It seems that the two scholars were not in correspondence with each other despite their quite similar interests. In any case, no such material is found in Paneth's writings in the archive of the MPG in Berlin.

chemistry as a "systematic art." In his *Critique of Judgment* (1st edition 1790) he characterizes the concept of "art" in three distinctions (§ 43: *Art in general*). The first of these distinctions is that of art and nature, in the sense of work and effect. Although the honeycomb structure in a beehive may occasionally be described as "artful," it is a natural phenomenon. The second distinction is between ability and knowledge, or practical and theoretical capacity, or also technology and theory (with the examples of mensuration and geometry).[31] Kant also feels that one can distinguish "art as human skill", from science, like ability from knowledge. In so doing he ushers the systematic action that is so decisive in chemistry (but not only there) to a place behind the formation of theory. The third and final distinction is between art and handicraft. In other words, that is the distinction between "genuine" art and "mere" handicraft. The expression "systematic art" from *Metaphysical Foundations of Natural Science* best fits the second of Kant's later characterizations. In addition to a descriptive theory of nature, which while including a static section and a dynamic section nonetheless integrates the two systematically, Kant's fundamental two-part system provides for the realm of natural science, which encompasses both the improper and proper disciplines. The result is a plausible hierarchy beginning with more static descriptivism and attaining the highest level with theoretical mathematical problem-solving. As was mentioned previously, Kant views chemistry as an improper science because its basis does not exhibit any mathematically determinable apodictic truth.[32]

The second variant of the interpretation of the art in *Critique of Judgment* corresponds to several descriptions by scholars in the field of chemistry.[33] Andreas Libavius, as quoted above, made mention of technical ("artificial") purification

---

31 With respect to the second distinction, Kant offers an interesting comment: "[...] what one *can* do the moment one only *knows* what is to be done, hence without anything more than sufficient knowledge of the desired result, is not called art. (Kant 1790, p. 163 or A172–173; original author's emphasis). Probably thinking of the ideal form of a shoe, he then mentions a shoemaker who will not be able to make this shoe.

32 In his book *Systematicity*, Paul Hoyningen-Huene develops a general approach to characterizing scientific nature. He identifies nine essential aspects that taken together distinguish scientific knowledge from other knowledge (Hoyningen-Huene 2013, p. 27): descriptions, explanations, predictions, defense of claims to knowledge, critical discourse, epistemic linkage, ideal of completeness, generation of knowledge, exposition of knowledge. In his comparison to the Kantian position, Hoyningen-Huene finds that practically all these aspects of his systematicity theory are implicitly advocated by Kant in the case of both the proper and improper sciences (Hoyningen-Huene 2013, pp. 158–159). The adjective in Kant's description "systematic art" thus makes it clear that this is in no way a disparagement. As I attempt to show, one can even interpret this description positively.

33 Cf. the statements by Boerhaave and Macquer quoted in the introductory chapter. There one may find a statement by historian of chemistry Fierz-David who feels that modern chemistry (current as of 1952) could still be regarded as a systematic art.

and manufacturing processes concerning specific epistemic objects. The "purification" of specimens of substances in the laboratory does not require any theory in the strict sense, although it does require a basic premise as to what constitutes a pure substance. The criterion for the success of such an operation is a practical one comparable to the successes in the art of cooking or the art of medicine. Highly developed specific processes have also been designed to obtain those substances that since the late 18th century have been referred to as elements; initially these processes functioned entirely without a theoretical superstructure. Many chemical elements have been demonstrated by systematic laboratory work but without theoretical explanations or predictions.[34] One example of this is the electrolytic dissociation of water with which it was possible to explain that this substance is not an element in the modern sense but "consists of" two different elements. With the specimens of these elements obtained in this manner, one acquires knowledge about their nature and a web of relational knowledge unfolds. It is in any case futile to hope that the import of every tiny bit of physical theory is a step closer to the supposedly ennobling label of "proper science," for chemistry is fundamentally different from physics in its pursuit of knowledge. Even after having achieved higher degrees of mathematization in individual subfields, its main task remains the realization of substances and substance relationships. On the one hand, Kant's adjective "systematic" teaches us that chemistry is more than merely a handicraft. On the other hand, no science, particularly chemistry, is possible without experimental "skill."

Aside from the functional goal of improving living conditions by manufacturing and obtaining new fuels, foodstuffs, materials, and active ingredients, there are also fundamental aspects like the growth of knowledge by expanding reality.[35] In chemistry substances might be manufactured for their own sake, although this is often represented differently in the "official" literature.[36] Chemistry "makes" its own epistemic objects, and even in analytics it cannot do without synthesis, that is substance formation. This means that Kant is correct. Chemistry is indeed a sys-

34 Here I remind the reader of the passage in the writings of František Wald quoted above in which he denounces the naive notion of the "preexistence" of elements and compounds and notes that substances, like skeletons and mummies, are in reality preparations. Cf. Wald (1902a, p. 16).

35 Here I have borrowed part of the title of publications regarding the philosophy of experiments by Michael Heidelberger (*Die Erweiterung der Wirklichkeit im Experiment*, 1997 and 1998) discussed previously.

36 It has become common practice, and institutions providing work and third-party funding have come to expect that descriptions of synthetic projects also include descriptions of the possible applications of the products desired or obtained. This can occasionally lead to grotesque expositions whose contents have no chance of ever being realized.

tematic art, namely the art of understanding substances by manufacturing, measurement and determination. Not only does it interpret the world, it also changes it, scientifically and technically, locally and globally, for better and for worse.

Fig. 27: The confident chemist.

12 References

If not specified otherwise, historical details are taken from the following sources:

Autorenkollektiv (1988) *Lexikon bedeutender Chemiker.* Thun: Harri Deutsch.
Autorenkollektiv (1989) *ABC Geschichte der Chemie.* Leipzig: VEB Deutscher Verlag für Grundstoffindustrie.
All cited letters and other sources referring to Friedrich Paneth are in the Archives of the Max-Planck-Gesellschaft, Berlin-Dahlem (Abteilung III, Repositur 45). I thank Mr. Hoffmann for his support there.
Details about life and work of František Wald are from the archives of the Wald family. I thank Jiři Wald (Prague) for the generous copy of the electronic archives.
Information regarding nominations for Nobelprizes are from www.Nobelprize.org.

Abegg, R. (1904) Die Valenz und das periodische System. Versuch einer Theorie der Molekularverbindungen. *Zeitschrift für anorganische und allgemeine Chemie* 39, 330–380.
Abegg, R., Bodländer, G. (1899) Die Elektroaffinität, ein neues Prinzip der chemischen Systematik. *Zeitschrift für anorganische und allgemeine Chemie* 20, 453–499.
Amann, A., Gans, W. (1989) Die Theoretische Chemie auf dem Weg zu einer Theorie der Chemie. *Angewandte Chemie* 101, 277–285.
Arabatzis, T. (2006) *Representing Electrons.* Chicago/London: The University of Chicago Press.
Aristotle (1952) *Meteorologica.* With an English Translation by H.D.P. Lee. Cambridge, Massachusetts/ London: Harvard University Press/ William Heinemann Ltd.
Arrhenius, S. (1889) Über die Reaktionsgeschwindigkeit bei der Inversion von Rohrzucker durch Säuren. *Zeitschrift für Physikalische Chemie* 4, 226–248.
Arrhenius, S. (1906) *Theorien der Chemie.* Leipzig: Akademische Verlagsgesellschaft.
Arriaga, J.A.J., Fortin, S., Lombardi, O. (2019) A New Chapter in the Problem of the Reduction of Chemistry to Physics: The Quantum Theory of Atoms in Molecules. *Foundations of Chemistry* 21, 125–136.
Aston, F.W. (1942) *Mass Spectra and Isotopes.* 2nd edition. London: Edward Arnold & Co.
Atkins, P. (1993) *Quanten. Begriffe und Konzepte für Chemiker.* Weinheim: VCH.
Atkins, P. (2015) Chemistry – A Very Short Introduction. Oxford: Oxford University Press.
Atkins, P., de Paula, J. (2013) *Elements of Physical Chemistry.* 6th edition. Oxford: Oxford University Press.
Autor collective (1988) *Philosophie und Naturwissenschaften.* Berlin: VEB Deutscher Verlag der Wissenschaften.
Bachelard, G. (1940/1980) *Die Philosophie des Nein.* Frankfurt am Main: Suhrkamp.
Bachelard, G. (1953/1972) *Le Matérialisme rationnel.* Paris: Presses Universitaires de France.
Bachelard, G. (1993) *Epistemologie.* Frankfurt am Main: Fischer (Original 1971).
Bader, R.F.W., Matta, C.F. (2013) Atoms in Molecules as Non-overlapping, Bounded, Space-Filling Open Quantum Systems. *Foundations of Chemistry* 15, 253–276.
Bader, R.F.W., Popelier, P.L., Keith, T.A. (1994) Die theoretische Definition einer funktionellen Gruppe und das Paradigma des Molekülorbitals. *Angewandte Chemie* 106, 647–659.
Baird, D., Scerri, E., McIntyre, L., Eds. (2006) *Philosophy of Chemistry – Synthesis of a New Discipline.* Dordrecht: Springer.

https://doi.org/10.1515/9783111401041-015

Baucke, F.G.K. (1994) The Modern Understanding of the Glass Electrode Response. *Fresenius Zeitschrift für Analytische Chemie* 349, 582–596.

Baumgärtel, E., Lehmann, R., Lichtner, E. (1972) *Grundlagen der Allgemeinen Chemie – Für Hochschulingenieure und Pädagogen.* Leipzig: Geest & Portig.

Beaney, M. (2021) Analysis. *The Stanford Encyclopedia of Philosophy* (Summer 2021 Edition), Edward N. Zalta (ed.), https://plato.stanford.edu/archives/sum2021/entries/analysis/>. Last access: 01.04.24

Beauchamp, T.L., Childress, J.F. (2009) Principles of Biomedical Ethics. 6th edition. Oxford/New York: Oxford University Press.

Bennett, M.R., Hacker, P.M.S. (2003) *Philosophical Foundations of Neuroscience.* Malden: Blackwell.

Bensaude-Vincent, B. (2012) Gaston Bachelard (1884–1962). In Hendry, R.F. et al., Eds., *Handbook of the Philosophy of Science.* Vol. 6: *Philosophy of Chemistry,* 141–150. Amsterdam: Elsevier.

Bensaude-Vincent, B., Simon, J. (2008) *Chemistry – The Impure Science.* London: Imperial College Press.

Bensaude-Vincent, B., Stengers, I. (1996) *A History of Chemistry.* Cambridge/London: Harvard University Press.

Berthelot, M. (1860) *Chimie Organique fondée sur la Synthèse.* Paris: Mallet-Bachelier.

Berthelot, M. (1876) *La Synthèse Chimique.* Paris: Germer Baillière.

Berthelot, M. (1877) *Die chemische Synthese.* Leipzig: F.A. Brockhaus.

Berthelot, M. (1909) *Die Chemie im Altertum und im Mittelalter.* Transl. by Emma Kalliwoda. Leipzig/Vienna: Franz Deuticke.

Berthollet, C.L. (1803) *Essai de Statique Chimique.* Paris: L'Imprimerie de Demonville et Soeurs.

Berzelius, J.J. (1836) Einige Ideen über die bei der Bildung organischer Verbindungen in der lebenden Natur wirksame, aber bisher nicht bemerkte Kraft. *Jahresbericht über die Fortschritte der physischen Wissenschaften* 15, 237–245.

Bhushan, N. (2006) Are Chemical Kinds Natural Kinds? In Baird, D., Scerri, E., McIntyre, L., Eds., *Synthesis of a New Discipline*, 327–336. Dordrecht: Springer.

Bird, A., Tobin, E. (2024) Natural Kinds. *The Stanford Encyclopedia of Philosophy* (Spring 2024 Edition), Edward N. Zalta & Uri Nodelman (eds.), URL = <https://plato.stanford.edu/archives/spr2024/entries/natural-kinds/>. Last access: 01.04.24.

Bodenstein, M. (1894) Über die Zersetzung des Jodwasserstoffgases in der Hitze. *Zeitschrift für Physikalische Chemie* 13, 56–127.

Boerhaave, H. (1755) *Anfangsgründe der Chemie.* Hannover (Latin original Leiden 1732).

Bogaard, P.A. (2003) How G.N. Lewis Reset the Terms of the Dialogue between Chemistry and Physics. In Earley, J., Ed., *Chemical Explanation*, 307–312. New York: The New York Academy of Sciences.

Böhme, G. (1980) *Alternativen der Wissenschaft.* Frankfurt am Main: Suhrkamp.

Böhme, G., Böhme, H. (2004) *Feuer Wasser Erde Luft – Eine Kulturgeschichte der Elemente.* Munich: C.H. Beck.

Böhme, G., van den Daele, W. (1977) Erfahrung als Programm – Über Strukturen vorparadigmatischer Wissenschaft. In Böhme, G., van den Daele, W., Krohn, W., *Experimentelle Philosophie*, 183–236. Frankfurt am Main: Suhrkamp.

Böhme, G., van den Daele, W., Krohn, W. (1978) Die Verwissenschaftlichung von Technologie. In Böhme, G., van den Daele, W., Hohlfeld, R., Krohn, W., Schäfer, W., Spengler, T., Eds., *Starnberger Studien* I, *Die gesellschaftliche Orientierung des wissenschaftlichen Fortschritts*, 339–375. Frankfurt am Main: Suhrkamp.

Boyce, C. (2019) Mendeleev's Elemental Ontology and Its Philosophical Renditions in German and English. *HYLE – International Journal for Philosophy of Chemistry* 25, 49 – 70.

Boyd, R. (1973) Realism, Underdetermination, and a Causal Theory of Evidence. *Noûs* 7, 1 – 12.

Boyd, R., Gasper, P., Trout, J.D., Eds. (1991) *The Philosophy of Science.* Cambridge: MIT Press.

Boyd, W.C., Boyd, R. (2019) Natural Kinds and Ceteris Paratis Generalizations: In Praise of Hunches. *HYLE – International Journal for Philosophy of Chemistry* 25, 21 – 47.

Bradley, J. (1955) On the Operational Interpretation of Classical Chemistry. *The British Journal for the Philosophy of Science* 6, 32 – 42.

Bradley, J. (1971) *Mach's Philosophy of Science.* London: Athlone Press.

Brock, W. (2004) Why Helium Ends in "-ium"? *Journal of Chemical Education* 81, 944.

Brooke, J.H. (1971) Organic Synthesis and the Unification of Chemistry: A Reappraisal. *The British Journal for the History of Science* 5, 363 – 392.

Brush, S.G. (1987) *Die Temperatur der Geschichte.* Braunschweig/Wiesbaden: Friedrich Vieweg & Sohn.

Buchdahl, G. (1959) Sources of Scepticism in Atomic Theory. *The British Journal for the Philosophy of Science* 10, 120 – 134.

Buchner, G. (1912) *Angewandte Ionenlehre.* Munich: J.F. Lehmann.

Buswell, A.M. (1922) Boutron Boudet Soap Solution. *Journal of the American Water Works Association* 9, 892 – 896.

Cammann, K. (1992) Analytical Chemistry – Today's Definition and Interpretation. *Fresenius' Journal of Analytical Chemistry* 343, 812 – 813.

Carrier, M. (1986) Die begriffliche Entwicklung der Affinitätstheorie im 18. Jahrhundert Newtons Traum – und was daraus wurde. *Archive for History of Exact Sciences* 36, 327 – 389.

Carrier, M. (1990) Kants Theorie der Materie und ihre Wirkung auf die zeitgenössische Chemie. *Kant-Studien* 82, 170 – 210.

Carrier, M. (2011) Underdetermination as an Epistemological Test Tube: Expounding Hidden Values of the Scientific Community. *Synthese* 180, 189 – 204.

Chadwick, J. (1932) The Existence of a Neutron. *Proceedings of the Royal Society of London* A 136, 692 – 708.

Chang, H. (1995) The Quantum Counter-Revolution: Internal Conflicts in Scientific Change. *Studies in the History and Philosophy of Modern Physics* 26, 121 – 136.

Chang, H. (1997) On the Applicability of the Quantum Measurement Formalism. *Erkenntnis* 46, 143 – 163.

Chang, H. (2004) *Inventing Temperature – Measurement and Scientific Progress.* Oxford: Oxford University Press.

Chang, H. (2011) The Persistence of Epistemic Objects through Scientific Change. *Erkenntnis* 75, 413 – 429.

Chang, H. (2012) *Is Water H_2O? Evidence, Realism and Pluralism.* Dordrecht: Springer.

Chang, H. (2016) The Rising of Chemical Natural Kinds through Epistemic Iteration. In Kendig, C., Ed., *Natural Kinds and Classification in Scientific Practice*, 33 – 46. London: Routledge.

Chang, H. (2017) What History Tells Us about the Distinct Nature of Chemistry. *Ambix* 64, 360 – 374.

Chang, H. (2022) *Realism for Realistic People. A New Pragmatist Philosophy of Science.* Cambridge: Cambridge University Press.

Christie, M. (2000) *The Ozone Layer – A Philosophy of Science Perspective.* Cambridge: Cambridge University Press.

Cornelius, H. (1916) *Transcendentale Systematik.* Munich: Ernst Reinhardt.

Cornelius, H. (1921) *Einleitung in die Philosophie.* 2nd edition. Leipzig/Berlin: B.G. Teubner.

Cornelius, H. (1931) Zur Kritik der wissenschaftlichen Grundbegriffe. *Erkenntnis* 2, 191–218.
Coulson, C.A. (1952) Quantum Theory, Theory of Molecular Structure and Valence. *Annual Reviews of Physical Chemistry* 3, 1–18.
Cremer, M. (1906) Über die Ursache der elektromotorischen Eigenschaften der Gewebe, zugleich ein Beitrag zur Lehre von den polyphasischen Elektrolytketten. *Zeitschrift für Biologie* 47, 562–608.
Cremer, M. (1928) Ursache der elektrischen Erscheinungen. In *Handbuch der normalen und pathologischen Physiologie*, Part 2, 999–1053. Berlin: Julius Springer.
Crutzen, P.J. (1970) The Influence of Nitrogen Oxides on the Atmospheric Ozone Content. *Quarterly Journal of the Royal Meteorological Society* 96, 320–325.
Crutzen, P.J., Oppenheimer, M. (2008) Learning about Ozone Depletion. *Climate Change* 89, 143–154.
Dainton, F.S. (1975) Faraday Lecture. The Electron as a Chemical Entity. Chemical Society Reviews 3, 323–362.
Danzer, K. (1992) Analytical Chemistry – Today's Definition and Interpretation. *Fresenius' Journal of Analytical Chemistry* 343, 827–828.
Deltete, R. (2012) Friedrich Wilhelm Ostwald (1853–1932). In Hendry, R.F. et al., Eds., *Handbook of the Philosophy of Science.* Vol. 6: *Philosophy of Chemistry*, 101–111. Amsterdam: Elsevier.
Dingle, H., Martin, G.R. (1964) *Chemistry and Beyond.* New York: Interscience.
Dingler, H. (1928) *Das Experiment – sein Wesen und seine Geschichte.* Munich: Ernst Reinhardt.
Dobson, G.M.B. (1968) Forty Years' Research on Atmospheric Ozone at Oxford: A History. *Applied Optics* 7, 387–405.
Dole, M. (1941) *The Glass Electrode – Methods, Applications, and Theory.* New York: John Wiley & Sons.
Druce, G. (1944) *Two Czech Chemists. Bohuslav Brauner (1855–1935)/Frantisek Wald (1861–1930).* London: The New Europe Publishing Company.
Düllmann, Ch.E., Brüchle, W., Dressler, R., Eberhardt, K., Eichler, B., Eichler, R., Gäggeler, H.W., Ginter, T.N., Glaus, F., Gregorich, K.E., Hoffman, D.C., Jäger, E., Jost, D.T., Kirbach, U.W., Lee, D.M., Nitsche, H., Patin, J.B., Pershina, V., Piguet, D., Qin, Z., Schädel, M., Schausten, B., Schimpf, E., Schött, H.-J., Soverna, S., Sudowe, R., Thörle, P., Timokhin, S.N., Trautmann, N., Türler, A., Vahle, A., Wirth, G., Yakushev, A.B., Zielinski, P.M. (2002) Chemical Investigation of Hassium (Element 108). *Nature* 418, 859–861.
Dupré, J. (1995) *The Disorder of Things.* Cambridge/London: Harvard University Press.
Earley, J., Ed. (2003) Chemical Explanation. *Annals of the New York Academy of Sciences* 988. New York: New York Academy of Sciences.
Earley, J. (2008) How Philosophy of Mind Needs Philosophy of Chemistry. *HYLE – International Journal for Philosophy of Chemistry* 14, 1–26.
Earley, J. (2009) How Chemistry Shifts Horizons: Element, Substance, and the Essential. *Foundations of Chemistry* 11, 65–77.
Ehrenfeld, R. (1906) *Grundriß einer Entwicklungsgeschichte der chemischen Atomistik.* Heidelberg: Carl Winter's Universitätsbuchhandlung.
Eisenman, G., Ed. (1967) *Glass Electrodes for Hydrogen and Other Cations.* New York: Marcel Dekker.
Emeléus, H.J. (1960) Friedrich Adolf Paneth. *Biographical Memoirs of Fellows of the Royal Society* 6, 227–239.
Ertl, G. (1990) Elementarschritte bei der heterogenen Katalyse. *Angewandte Chemie* 102, 1258–1266.
Ertl, G., Soentgen, J. (2015) *N – Stickstoff – ein Element schreibt Weltgeschichte.* Munich: oekom.
Fabian, P. (1987) Atmosphäre und Umwelt. 2nd edition. Berlin/Heidelberg/New York: Springer.
Fabian, P. (1995) Atmosphäre und Umwelt. 4th edition. Berlin/Heidelberg/New York: Springer.
Fajans, K. (1919) *Radioaktivität.* Braunschweig: Friedrich Vieweg & Sohn.

Falk, K.G., Nelson, J.M. (1910) The Electron Conception of Valence. *Journal of the American Chemical Society* 32, 1637 – 1654.
Feyerabend, P. (1975) *Against Method.* London: Verso.
Fierz-David, H.E. (1952) *Die Entwicklungsgeschichte der Chemie.* 2nd edition. Basel: Birkhäuser.
Fieser, L., Fieser, M. (1968) *Organische Chemie.* 2nd edition. Weinheim: Verlag Chemie.
Finston, H.L., Rychtman, A.C. (1982) *New View of Current Acid-Base Theories.* New York: John Wiley and Sons.
Fortin, S., Lombardi, O., Martínez González, J.C. (2016) Isomerism and Decoherence. *Foundations of Chemistry* 18, 225 – 240.
French, S. (2011) Metaphysical Underdetermination: Why Worry? *Synthese* 180, 205 – 221.
Fresenius, C.R. (1866) *Anleitung zur qualitativen chemischen Analyse.* 12th edition. Braunschweig: Friedrich Vieweg & Sohn.
Freund, I. (1904) *The Study of Chemical Composition.* Cambridge: At the University Press.
Frevert, M. (2014) *Wissenschaft der Synthese – Ein Ansatz zur wissenschaftsphilosophischen Bestimmung der Chemie.* Kassel: Kassel University Press.
Friedman, M. (1992) *Kant and the Exact Sciences.* Cambridge/London: Harvard University Press.
Friege, H., Engelhardt, C., Henseling, K.O., Eds. (1998) *Das Management von Stoffströmen.* Berlin: Springer.
Gavroglu, K., Simões, A. (2012) *Neither Physics nor Chemistry. A History of Quantum Chemistry.* Cambridge: The MIT Press.
Glasstone, S., Laidler, K., Eyring, H. (1941) *The Theory of Rate Processes.* New York/London: McGraw-Hill.
Gmelin, L. (1843) *Handbuch der Chemie.* Vol. 1. 1st edition. Heidelberg: K. Winter.
Gomberg, M. (1900) An Instance of Trivalent Carbon: Triphenylmethyl. *Journal of the American Chemical Society* 22, 757 – 771.
Goodman, N. (1988) *Tatsache, Fiktion, Voraussage.* Frankfurt am Main.: Suhrkamp.
Görs, B., Psarros, N., Ziche, P., Eds. (2005) *Wilhelm Ostwald at the Croccroads between Chemistry, Philosophy and Media Culture.* Leipzig: Leipziger Universitätsverlag.
Graham, Th. (1869) On the Relation of Hydrogen to Palladium. *Proceedings of the Royal Society of London* 17, 212 – 220.
Guldberg, C.M., Waage, P. (1879) Ueber die chemische Affinität. *Journal für praktische Chemie* 19, 69 – 114.
Guntau, M., Wendt, H., Eds. (1967) *Naturforschung und Weltbild. Eine Einführung in philosophische Probleme der modernen Naturwissenschaften.* Berlin: VEB Deutscher Verlag der Wissenschaften.
Haber, F., Klemensiewicz, Z. (1909) Über elektrische Phasengrenzkräfte. *Zeitschrift für physikalische Chemie* 67, 385 – 432.
Hacker, P.M.S. (1979) Substance: The Constitution of Reality. *Midwest Studies in Philosophy* 4, 239 – 261.
Hacker, P.M.S. (2004) Substance: Things and Stuffs. *Supplement to the Proceedings of the Aristotelian Society* 78, 41 – 63.
Hacking, I. (1982) Experimentation and Scientific Realism. *Philosophical Topics* 13, 154 – 172.
Hacking, I. (1983) *Representing and Intervening.* Cambridge: Cambridge University Press.
Hacking, I. (1991) A Tradition of Natural Kinds. *Philosophical Studies* 61, 109 – 126.
Haller, R., Stadler, F., Eds. (1988) *Ernst Mach. Werk und Wirkung.* Vienna: Hölder-Pichler-Tempsky.
Hanekamp, G. (1997) *Protochemie – Vom Stoff zur Valenz.* Würzburg: Königshausen & Neumann.
Harré, R. (1986) *Varieties of Realism – a Rationale for the Natural Sciences.* Oxford: Basil Blackwell.

Harré, R. (2009) Trope Theory and the Ontology of Chemistry. *Foundations of Chemistry* 11, 93 – 103.

Harré, R. (2014) New Tools for Philosophy of Chemistry. *HYLE – International Journal for Philosophy of Chemistry* 20, 77 – 91.

Hartmann, D., Janich, P. (1996) Methodischer Kulturalismus. In Hartmann, D., Janich, P., Eds., *Methodischer Kulturalismus.* Frankfurt am Main: Suhrkamp, 9 – 69.

Hartmann, N. (1964) *Der Aufbau der realen Welt.* 3rd edition. Berlin: Walter De Gruyter.

Haupt, B. (1987) *Deutschsprachige Chemielehrbücher* (1775 – 1850). Stuttgart: Deutscher Apotheker Verlag.

Havemann, R. (1957) *Einführung in die chemische Thermodynamik.* Berlin: VEB Deutscher Verlag der Wissenschaften.

Havemann, R. (1964) *Dialektik ohne Dogma? – Naturwissenschaft und Weltanschauung.* Hamburg: Rowohlt.

Headley, C. (2020) Philosophy, Natural Kinds, Microstructuralism, and the (Mis)use of Chemical Examples: Intimacy versus Integrity as Orientations towards Chemical Practice. *Foundations of Chemistry* 22, 489 – 500.

Heidelberger, M. (1997) Die Erweiterung der Wirklichkeit im Experiment. *ZiF: Mitteilungen* 2, 1 – 11.

Heidelberger, M. (2003) Theory-Ladenness and Scientific Instruments in Experimentation. In Radde, H., Ed., *The Philosophy of Scientific Experimentation*, 138 – 151. Pittsburgh, PA: University of Pittsburgh Press.

Heisenberg, W. (1989) *Ordnung der Wirklichkeit.* Munich/Zürich: Piper.

Heitler, W. (1961) *Elementare Wellenmechanik. Mit Anwendungen auf die Quantenchemie.* Braunschweig: Friedrich Vieweg & Sohn.

Heitler, W., London, F. (1927) Wechselwirkung neutraler Atome und homöopolare Bindung nach der Quantenmechanik. *Zeitschrift für Physik* 44, 455 – 472.

Hellmann, H. (1937/2015) *Einführung in die Quantenchemie. Mit biografischen Notizen von Hans Hellmann jr.* Ed. D. Andrae. Berlin/Heidelberg: Springer Spektrum.

Hempel, C.G. (1974) *Philosophie der Naturwissenschaften.* Munich: Deutscher Taschenbuchverlag.

Hempel, C.G. (1977) *Aspekte wissenschaftlicher Erklärung.* Berlin/New York: De Gruyter.

Hendry, R.F. (2006) Is There a Downward Causation in Chemistry? In Baird, D., Scerri, E., McIntyre, L., Eds., *Philosophy of Chemistry*, 173 – 189. Dordrecht: Springer.

Hendry, R.F. (2012) Charles Coulson (1910 – 1974). In Hendry, R.F. et al., Eds., *Handbook of the Philosophy of Science.* Vol. 6: *Philosophy of Chemistry*, 159 – 164. Amsterdam: Elsevier.

Hendry, R.F. (2012) Elements. In Hendry, R.F. et al., Eds., *Handbook of the Philosophy of Science.* Vol. 6: *Philosophy of Chemistry*, 255 – 269. Amsterdam: Elsevier.

Hendry, R.F. (2017) Prospects for Strong Emergence in Chemistry. In Paoletti, M.P., Orilia, F, Eds., *Philosophical and Scientific Perspectives on Downward Causation.* Routledge Studies in Contemporary Philosophy 91, 146 – 163. New York: Routledge.

Hendry, R.F. (2021) Structure, Scale and Emergence. *Studies in History and Philosophy of Science* 85, 44 – 53.

Hendry, R.F., Needham, P. Woody, A., Eds. (2012) *Handbook of the Philosophy of Science.* Vol. 6: *Philosophy of Chemistry.* Amsterdam: Elsevier.

Hentschel, K. (2000) Historiographische Anmerkungen zum Verhältnis von Experiment, Instrumentation und Theorie. In Meinel, C., Ed., *Instrument-Experiment*, 13 – 51. Berlin/Diepholz: Verlag für Geschichte der Naturwissenschaften und der Technik.

Hettema, H. (2012) The Unity of Chemistry and Physics: Absolute Reaction Rate Theory. *HYLE – International Journal for Philosophy of Chemistry* 18, 145 – 173.

Hevesy, G. de (1964) Some Applications of Isotopic Indicators. *Nobel Lectures, Chemistry 12, 1942 – 1962*, 9 – 41. Amsterdam: Elsevier.
Hijmans, S. (2020) Chemical Elements and Chemical Substances – Rethinking Paneth's Distinction. In Scerri, E., Ghibaudi, E., Eds., *What Is a Chemical Element?*, 241 – 256. Oxford: Oxford University Press.
Höffe, O. (2011) *Kants Kritik der reinen Vernunft – Die Grundlegung der modernen Philosophie.* Munich: C.H. Beck.
Hoffmann, D. (2005) Between Autonomy and Accomodation: The German Physical Society during the Third Reich. *Physics in Perspective* 7, 293 – 329.
Hoffmann, R. (1995) *The Same and Not the Same.* New York: Columbia University Press.
Hoffmann, R. (2007) What Might Philosophy of Science Look Like If Chemists Built It? *Synthese* 155, 321 – 336.
Holleman, A.F., Wiberg, E. (1964) *Lehrbuch der Anorganischen Chemie.* Berlin: De Gruyter.
Hooykaas, R. (1958) The Concepts of Individual and Species in Chemistry. *Centaurus* 5, 307 – 322.
Hoyningen-Huene, P. (2013) *Systematicity – The Nature of Science.* Oxford: Oxford University Press.
Hu, S.-X, Li, W.-L, Lu, J.-B., Bao, J.L., Yu, H.S., Truhlar, D.G., Gibson, J.K., Marçalo, J. Zhou, M., Riedel, S., Schwarz, W.H.E., Li, J. (2018) Über Oxidationszahl-Obergrenzen in der Chemie. *Angewandte Chemie* 130, 3297 – 3300.
Huizenga, J.R. (1993) *Cold Fusion – The Scientific Fiasco of the Century.* Oxford: Oxford University Press.
Huizinga, J. (1956) *Homo Ludens. Vom Ursprung der Kultur im Spiel.* Hamburg: Rowohlt.
Hund, F. (1927) Zur Deutung der Molekelspektren III. *Zeitschrift für Physik* 43, 805 – 826.
Hund, F. (1967) *Geschichte der Quantentheorie.* Mannheim: Bibliographisches Institut.
Hüttig, G. (1914) Der Begriff der "chemischen Konstitution" vom Standpunkte der Mannigfaltigkeitslehre. *Annalen der Naturphilosophie* 13, 260 – 274.
Ingold, C.K. (1934) Principles of an Electronic Theory of Organic Reactions. *Chemical Reviews* 15, 225 – 274.
Jackson, C.M. (2015) The "Wonderful Properties of Glass": Liebig's Kaliapparat and the Practise of Chemistry in Glass. *Isis* 106, 43 – 69.
Jackson, C.M. (2023) Molecular World – Making Modern Chemistry. Cambridge/Massachusetts: MIT Press.
Jacob, C. (2001) Analysis and Synthesis. *HYLE – International Journal for Philosophy of Chemistry* 7, 31 – 50.
Janich, P. (1980) Chemie. In Mittelstraß, J., Ed., *Enzyklopädie Philosophie und Wissen-schaftsheorie.* Mannheim: Bibliographisches Institut.
Janich, P. (1992) *Grenzen der Naturwissenschaft.* Munich: C.H. Beck.
Janich, P., Ed. (1994a) *Philosophische Perspektiven der Chemie.* Mannheim: BI Wissenschaftsverlag.
Janich, P. (1994b) Vorwort des Herausgebers. In Janich, P., Ed., *Philosophische Perspektiven der Chemie*, 1 – 9. Mannheim: BI Wissenschaftsverlag.
Janich, P. (1996) Natürlich künstlich. Philosophische Reflexionen zum Naturbegriff der Chemie. In Janich, P., Rüchardt, Chr., Eds., *Natürlich, technisch, chemisch – Verhältnisse zur Natur am Beispiel der Chemie*, 53 – 79. Berlin/New York: De Gruyter.
Janich, P. (2006) *Was ist Information?* Frankfurt am Main: Suhrkamp.
Janich, P., Psarros, N., Eds. (1996) *Die Sprache der Chemie.* Würzburg: Königshausen & Neumann.
Janich, P., Rüchardt Chr., Eds. (1996) *Natürlich, technisch, chemisch – Verhältnisse zur Natur am Beispiel der Chemie.* Berlin/New York: De Gruyter.
Jensen, W.B. (1978) The Lewis Acid-Base Definitions: A Status Report. *Chemical Reviews* 78, 1 – 22.

Jensen, W.B. (1984) Abegg, Lewis, Langmuir, and the Octet Rule. *Journal of Chemical Education* 61, 191 – 200.

Johnstone Stoney, G. (1881) On the Physical Units of Nature. *Philosophical Magazine* 11, 381 – 390.

Johnstone Stoney, G. (1894) On the "Electron" or Atom of Electricity. *Philosophical Magazine* 38, 418 – 420.

Kant, I. (1781/2003) Critique of Pure Reason. Transl. by J.M.D. Meiklejohn. Mineola, NY: Dover Publications, Inc

Kant, I. (1783) *Prolegomena zu einer jeden künftigen Metaphysik, die als Wissenschaft wird auftreten können.* Leipzig: Philipp Reclam jun.

Kant, I. (1786/1982) Metaphysische Anfangsgründe der Naturwissenschaft. In *Schriften zur Naturphilosophie* (Ed. W. Weischedel). Frankfurt am Main: Suhrkamp.

Kant, I. (1786/2004) *Metaphysical Foundations of Natural Science.* Transl. and ed. by M. Friedman. Cambridge: Cambridge University Press.

Kant, I. (1790/1952) The Critique of Judgement. Transl. by J.C. Meredith. Oxford: Clarendon Press. *Kritik der Urteilskraft.* Ed. W. Weischedel. Frankfurt am Main: Suhrkamp.

Kastner, C.W.G. (1814) *Einleitung in die neuere Chemie.* Halle/Berlin: Buchhandlung des Hallischen Waisenhauses.

Kauffmann, H. (1908) Elektronentheorie und Valenzlehre. *Physikalische Zeitschrift* 9, 311 – 314.

Kessel, N. (2013) Doriden von Ciba: Sleeping Pills, Pharmaceutical Marketing, and Thalidomide, 1955 – 1963. *History and Technology* 29, 153 – 168.

Kim, M.G. (2014) Stabilizing Chemical Reality: The Analytic-Synthetic Ideal of Chemical Species, *HYLE – International Journal for Philosophy of Chemistry* 20, 117 – 139.

Kirk, B. (1999) *Der Contergan-Fall: eine unvermeidbare Arzneimittelkatastrophe? Zur Geschichte des Arzneistoffs Thalidomid.* Stuttgart: Wissenschaftliche Verlagsgesellschaft.

Klaus, G., Buhr, M., Eds. (1974) *Philosophisches Wörterbuch*, Band 1 und 2. Westberlin: Verlag das Europäische Buch.

Klein, U. (1994) Origin of the Concept of Chemical Compound. *Science in Context* 7, 163 – 204.

Klein, U. (2008) A Historical Ontology of Material Substances: c. 1700 – 1830. In Ruthenberg, K., van Brakel, J., Eds., *Stuff – The Nature of Chemical Substances*, 21 – 42. Würzburg: Königshausen & Neumann.

Klein, U., Lefèvre, W. (2007) *Materials in Eighteenth-Century Science – A Historical Ontology.* Cambridge, MA.: MIT Press.

Knightley, P. (1979) *Suffer the Children – The Story of Thalidomide.* London: A. Deutsch.

Knoche, B., Blaschke, G. (1994) Investigations on the In Vitro Racemization of Thalidomide by High-Performance Liquid Chromatography. *Journal of Chromatography A* 666, 235 – 240.

Korte, F., Ed. (1987) *Lehrbuch der Ökologischen Chemie.* Stuttgart/New York: Georg Thieme.

Kragh, H. (2000) Conceptual Changes in Chemistry: The Notion of a Chemical Element, ca. 1900 – 1925. *Studies in History and Philosophy of Modern Physics* 31, 435 – 450.

Kratz, L. (1950) *Die Glaselektrode und ihre Anwendungen.* Frankfurt am Main: Dr. Dietrich Steinkopff.

Kuhn, Th. (1978) *Die Entstehung des Neuen.* Ed. L. Krüger. Frankfurt am Main: Suhrkamp.

Kunz, W., Keller, H., Mückter, H. (1956) N-Phthalyl-glutaminsäure-imid: Experimentelle Untersuchungen an einem neuen synthetischen Produkt mit sedativen Eigenschaften. *Arzneimittel-Forschung* 6, 426 – 430.

Kurnakov, N.S. (1914) Verbindung und chemisches Individuum. *Zeitschrift für Anorganische und Allgemeine Chemie* 88, 109 – 127.

Ladik, J. (1973) *Quantenchemie.* Stuttgart: Ferdinand Enke.

Ladyman, J. (2002) *Understanding Philosophy of Science.* London/New York: Routledge.

Laidler, K. (1970) *Reaktionskinetik I.* Mannheim/Vienna/Zürich: Bibliographisches Institut.

Laidler, K. (1986) The Development of Theories of Catalysis. *Archive for History of Exact Sciences* 35, 345 – 374.

Laidler, K. (1993) *The World of Physical Chemistry.* Oxford: Oxford University Press.

Laitinen, H.A., Ewing, G.W., Eds. (1977) *A History of Analytical Chemistry.* York, US: Maple Press Company.

Laitko, H. (1967) Philosophische Fragen der Chemie. Eine Einführung in die Problemsituation. In Guntau, M., Wendt, H., Eds., *Naturforschung und Weltbild. Eine Einführung in philosophische Probleme der modernen Naturwissenschaften,* 107 – 137. Berlin: VEB Deutscher Verlag der Wissenschaften.

Laitko, H. (1996) Chemie und Philosophie: Anmerkungen zur Entwicklung des Gebietes in der Geschichte der DDR. In Psarros, N., Ruthenberg, K., Schummer, J., Eds., *Philosophie der Chemie – Bestandsaufnahme und Ausblick,* 37 – 58. Würzburg: Königshausen & Neumann.

Laitko, H., Sprung, W.-D. (1970) *Chemie und Weltanschauung.* Schwerte/Ruhr: Hubert Freistühler.

Langmuir, I. (1919) The Arrangement of Electrons in Atoms and Molecules. *Journal of the American Chemical Society* 41, 868 – 934.

Laszlo, P. (1998) Chemical Analysis as Dematerialization. *HYLE – An International Journal for the Philosophy of Chemistry* 4, 29 – 38.

Latour, B. (1999) *Pandora's Hope – Essays on the Reality of Science Studies.* Cambridge: Harvard University Press.

Latour, B. (2000) *Die Hoffnung der Pandora.* Frankfurt am Main: Suhrkamp.

Lavoisier, A.-L. (1789/1965) *Elements of Chemistry.* Transl. by R. Kerr New York: Dover.

Lenhart-Schramm, N. (2016) *Die Haltung des Landes Nordrhein-Westfalen zu Contergan und den Folgen.* Doctoral thesis, Universität Münster.

Lenin, W.I. (1909/1985) *Materialismus und Empiriokritizismus.* Berlin: Dietz.

Lequan, M. (2000) *La chimie selon Kant.* Paris: Presses Universitaires De France

Lewis, G.N. (1916) The Atom and the Molecule. *Journal of the American Chemical Society* 38, 762 – 785.

Lewis, G.N. (1923) *Valence and the Structure of Atoms and Molecules.* New York: The Chemical Catalog Company.

Lewis, G.N. (1926) *The Anatomy of Science.* New Haven: Yale University Press.

Lewis, G.N. (1938) Acids and Bases. *Journal of the Franklin Institute* 226, 293 – 313.

Libavius, A. (1597/1964) *Die Alchemie des Andreas Libavius.* Weinheim: Verlag Chemie.

Lewowicz, L., Lombardi, O. (2013). Stuff versus Individuals. *Foundations of Chemistry* 15, 65-77.

Liebig, J. (1831) Über einen neuen Apparat zur Analyse organischer Körper und über die Zusammensetzung einiger organischen Substanzen. *Annalen der Physik und Chemie* 97, 1 – 43.

Liegener, Ch., Del Re, G. (1987) Chemistry vs. Physics, the Reduction Myth, and the Unity of Science. *Zeitschrift für allgemeine Wissenschaftstheorie* 18, 165 – 174.

Lockemann, G. (1950) *Geschichte der Chemie.* Erster Band. Berlin: Walter de Gruyter.

Lomonossow, M.W. (1741/1961) *Ausgewählte Schriften.* Vol. 1. Berlin: Akademie-Verlag.

Lorenz, K. (1977) *Die Rückseite des Spiegels.* Munich: Deutscher Taschenbuch Verlag.

Losee, J. (2001) *A Historical Introduction to the Philosophy of Science.* Oxford: Oxford University Press.

Lowry, Th. (1915) *Historical Introduction to Chemistry.* London: Macmillan.

Lowry, Th. (1934) Introductory Address: Free Radicals and Ions as Factors in Chemical Change. *Transactions of the Faraday Society* 30, 3 – 9.

Luder, W.F., Zuffanti, S. (1946) *The Electronic Theory of Acids and Bases.* London: John Wiley and Sons.

Lund, E.W. (1965) Guldberg and Waage and the Law of Mass Action. *Journal of Chemical Education* 42, 548–550.
Mach, E. (1872/1969) Die Geschichte und die Wurzel des Satzes von der Erhaltung der Arbeit. Reprinted in *Abhandlungen*, Ed. J. Thiele. Amsterdam: E.J. Bonset.
Mach, E. (1897) *Populärwissenschaftliche Vorlesungen.* Leipzig: Johann Ambrosius Barth.
Mach, E. (1900) *Die Principien der Wärmelehre.* 2nd edition. Leipzig: Johann Ambrosius Barth.
Mach, E. (1906) *Erkenntnis und Irrtum.* 2nd edition. Leipzig: Johann Ambrosius Barth.
Mach, E. (1922/1991) *Analyse der Empfindungen.* Reprint of the 9th edition. Darmstadt: Wissenschaftliche Buchgesellschaft.
Macquer, P.J. (1752) *Anfangsgründe der Theoretischen Chemie.* Leipzig: Lankisch. (French Original 1749).
Mahootian, F. (2020) Kant, Cassirer, and the Idea of Chemical Element. In Scerri, E., Ghibaudi, E., Eds., *What Is a Chemical Element?*, 143–166. Oxford: Oxford University Press.
Maier, H. (2015) *Chemiker im "Dritten Reich".* Weinheim: Wiley-VCH.
Martínez González, J.C., Fortin, S., Lombardi, O. (2019) Why Molecular Structure Cannot Be Strictly Reduced to Quantum Mechanics. *Foundations of Chemistry* 21, 31–45.
McIntyre, L. (2007) Emergence and reduction in chemistry: ontological or epistemological concepts? *Synthese* 155, 337-343.
Mehrtens, H., Richter, S., Eds. (1980) *Naturwissenschaft, Technik und NS-Ideologie.* Frankfurt am Main: Suhrkamp.
Meinel, Chr. (1983) Theory or Practice? The Eighteenth-Century Debate on the Scientific Status of Chemistry. *Ambix* 30, 121–132.
Meinel, Chr. (1984) "... die Chymie anwendbarer und gemeinnütziger zu machen" – Wissenschaftlicher Orientierungswandel in der Chemie des 18. Jahrhunderts. *Angewandte Chemie* 96, 326–334.
Meinel, Chr. (1986) Die Chemie an den Universitäten des 18. Jahrhunderts – Institutionalisierungsstufen und konzeptioneller Wandel. *Academiae Analecta* (Mededelingen van de Koninklijke Academie voor Wetenschappen, Letteren en Schone Kunsten van Beligië, Klasse der Wetenschappen) 48(4), 35–57.
Meinel, Chr., Ed. (2000) *Instrument – Experiment, Historische Studien.* Berlin/Diepholz: Verlag für Geschichte der Naturwissenschaften und der Technik.
Meixner, U. (1994) Von der Wissenschaft der Ontologie. *Logos* N.F. 1, 375–399.
Meyer, J.L. (1893) *Grundzüge der theoretischen Chemie.* 2nd edition. Leipzig: Breitkopf & Härtel.
Mets, A., Lõhkivi, E., Müürsepp, P., Eigi-Watkin, J., Eds. (2024) *Practical Realist Philosophy of Science.* Lanham etc.: Lexington Books.
Mittasch, A. (1938) *Katalyse und Determinismus. Ein Beitrag zur Philosophie der Chemie.* Berlin: Julius Springer.
Mittasch, A. (1948) *Von der Chemie zur Philosophie. Ausgewählte Schriften und Vorträge.* Ulm: J. Ebner.
Mittasch, A. (1951) *Wilhelm Ostwalds Auslösungslehre* (Sitzungsberichte der Heidelberger Akademie der Wissenschaften, Mathematisch-naturwissenschaftliche Klasse, 1. Abhandlung). Berlin: Springer.
Mittasch, A., Theis, E. (1932) *Von Davy und Döbereiner bis Deacon – Ein halbes Jahrhundert Grenzflächenkatalyse.* Berlin: Verlag Chemie.
Moran, B.T. (2005) *Distilling Knowledge.* Cambridge: Harvard University Press.
Morris, P., Ed. (2002) *From Classical to Modern Chemistry – The Instrumental Revolution.* London: Royal Society of Chemistry & Science Museum London.
Mosini, V. (2000) A Brief History of the Theory of Resonance and of Its Interpretation. *Studies in the History and Philosophy of Modern Physics* 31, 569–581.

Mulliken, R. (1934) A New Electroaffinity Scale, Together with Data on Valence States and on Valence Ionization Potentials and Electron Affinities. *Journal of Chemical Physics* 2, 782–793.

Murphy, D., Goldkind, S.F. (2005) The Regulatory and Ethical Challenges of Pediatric Research. In Santoro, M.A., Gorrie, T.M., Eds., *Ethics and the Pharmaceutical Industry*, 48–67. Cambridge: Cambridge University Press.

Needham, P. (2009) Reduction and Emergence: A Critique of Kim. *Philosophical Studies* 146, 93–116.

Needham, P. (2017) *Macroscopic Metaphysics – Middle-Sized Objects and Longish Processes.* Cham: Springer International.

Nernst, W. (1921) *Theoretische Chemie – Vom Standpunkte der Avogadroschen Regel und der Thermodynamik.* 8th–10th edition. Stuttgart: Ferdinand Enke.

Niedersen, U., Krug, H.-J., Pohlmann, L. (1992) Wilhelm Ostwald – Von der Reversibilität zur Irreversibilität. *Chemie in unserer Zeit* 26, 304–313.

Niese, S. (2007) Die Sternstunde der Naturwissenschaften: Die späteren Nobelpreislaureaten Spemann – Hevesy – Krebs – Staudinger und ihr politisch-soziales Umfeld. In Martin, B., Ed., *550 Jahre Albert-Ludwigsuniversität Freiburg – Festschrift.* Vol. 3, 259–279. Freiburg: Karl Alber.

Nordmann, A. (2006) From Metaphysics to Metachemistry. In Baird, D., Scerri, E., McIntyre, L., Eds., *Philosophy of Chemistry – Synthesis of a New Discipline*, 347–362. Dordrecht: Springer.

Nye, M.J. (1981) Berthelot's Anti-Atomism: A "Matter of Taste"? *Annals of Science* 38, 585–590.

Ostwald, W. (1877) *Volumchemische Studien über Affinität.* Master's thesis. Dorpat: H. Laakmann's Buchdruckerei u. Lithographie.

Ostwald, W. (1878) *Volumchemische und Optisch-Chemische Studien.* Doctoral thesis. Dorpat: Druck von H. Laakmann.

Ostwald, W. (1890a) Elektrische Eigenschaften halbdurchlässiger Scheidewände. *Zeitschrift für physikalische Chemie* 6, 71–82.

Ostwald, W. (1890b) *Grundriss der allgemeinen Chemie.* Leipzig: Wilhelm Engelmann.

Ostwald, W. (1894) *Die Wissenschaftlichen Grundlagen der Analytischen Chemie.* Leipzig: Wilhelm Engelmann.

Ostwald, W. (1899) *Grundriss der allgemeinen Chemie.* 3rd edition. Leipzig: Wilhelm Engelmann.

Ostwald, W. (1904) Elements and Compounds (Faraday Lecture). *Journal of the Chemical Society* 85, 506–522.

Ostwald, W. (1907) *Prinzipien der Chemie. Eine Einleitung in alle chemischen Lehrbücher.* Leipzig: Akademische Verlagsgesellschaft.

Ostwald, W. (1908) *Grundriss der Naturphilosophie.* Leipzig: Philipp Reclam jun.

Ostwald, W. (1909) The Fundamental Principles of Chemistry. An Introduction to All Textbooks of Chemistry. Authorized translation by Harry W. Morse. New York: Longmans, Green & Co.

Ostwald, W. (1910) *Die Forderung des Tages.* Leipzig: Akademische Verlagsgesellschaft.

Ostwald, W. (1917) *Grundriss der allgemeinen Chemie.* 5th edition. Dresden/Leipzig: Theodor Steinkopff.

Ostwald, W. (1927) *Lebenslinien. Eine Selbstbiographie.* Part 2. Berlin: Klasing.

Paneth, F. (1911) Über die Umlagerung des Chinidins und Cinchonidins durch Schwefelsäure. *Sitzungsberichte der Akademie der Wissenschaften Wien*, Abt. IIa, 120, 257–274.

Paneth, F. (1916) Über den Element- und Atombegriff in Chemie und Radiologie. *Zeitschrift für Physikalische Chemie* 91, 171–198.

Paneth, F. (1918) Elementtabelle und Atomtabelle. *Zeitschrift für Physikalische Chemie* 92, 677–684.

Paneth, F. (1920) Fajans, K., Radioaktivität und die neueste Entwicklung der Lehre von den chemischen Elementen (Buchbesprechung). *Die Naturwissenschaften* 8, 94–95.

Paneth, F. (1927a) Neuere Versuche über die Verwandlung von Wasserstoff in Helium. *Die Naturwissenschaften* 15, 379.

Paneth, F. (1927b) The Transmutation of Hydrogen into Helium. *Nature* 119, 706 – 707.

Paneth, F. (1928) *Radio-Elements as Indicators and Other Selected Topics in Inorganic Chemistry.* New York: McGraw-Hill.

Paneth, F. (1930) Die Entwicklung und der heutige Stand unserer Kenntnis über das natürliche System der Elemente. *Die Naturwissenschaften* 18, 964 – 976.

Paneth, F. (1931) Über die erkenntnistheoretische Stellung des chemischen Elementbegriffs. *Schriften der Königsberger Gelehrten Gesellschaft* 8(4), 101 – 125.

Paneth, F. (1934) The Use of Free Methyl and Ethyl in Chemical Synthesis. *Transactions of the Faraday Society* 30, 179 – 181.

Paneth, F. (1936a) Role of Chemistry in the Study of Atomic Transmutation. *Nature* 137, 560 – 562.

Paneth, F. (1936b/1964) Chemical Elements and Primordial Matter: Mendeleeff's View and the Present Position. In Dingle, H., Martin, G.R., Eds., *Chemistry and Beyond*, 53 – 72. New York: Interscience

Paneth, F. (1947) The Making of the Missing Chemical Elements. *Nature* 159, 8 – 10.

Paneth, F., Günther, P.L. (1933) Chemical Detection of Artificial Transmutation of Elements. *Nature* 131, 652 – 653.

Paneth, F., Herzfeld, K. (1931) Über Freies Methyl und Freies Äthyl. *Zeitschrift für Elektrochemie* 37, 577 – 582.

Paneth, F., Hofeditz, W. (1929) Über die Darstellung von freiem Methyl. *Berichte der Deutschen Chemischen Gesellschaft* 62, 1335 – 1347.

Paneth, F., Klever, E., Peters, K. (1927) Zur Frage der Existenz eines dreiatomigen Wasserstoffs. *Zeitschrift für Elektrochemie* 33, 102 – 107.

Paneth, F., Lautsch, W. (1931a) Über freie organische Radikale im Gaszustand, II. Mitteil.: Über die Darstellung von freiem Äthyl. *Berichte der Deutschen Chemischen Gesellschaft* 64, 2702 – 2707.

Paneth, F., Lautsch, W. (1931b) Über freie organische Radikale im Gaszustand, III. Mitteil.: Über den Mechanismus des Reagierens der freien Radikale. *Berichte der Deutschen Chemischen Gesellschaft* 64, 2708 – 2718.

Paneth, F., Loleit, H. (1935) Chemical Detection of Artificial Transmutation of Elements (Letters to the Editor). *Nature* 136, 950.

Paneth, F., Peters, K. (1926) Über die Verwandlung von Wasserstoff in Helium. *Die Naturwissenschaften* 14, 956 – 962.

Pauling, L. (1958) Quantum Theory and Chemistry. In Kockel, B., Macke, W., Papapetrou, A., Eds., *Max-Planck-Festschrift 1958.* Berlin: VEB Verlag der Wissenschaften.

Pauling, L. (1962) *Die Natur der Chemischen Bindung.* Weinheim: Verlag Chemie.

Perkins, M.J. (2000) *Radical Chemistry: The Fundamentals.* Oxford: Oxford University Press.

Pietsch, W. (2012) Defending Underdetermination or Why the Historical Perspective Makes a Difference. In de Regt, H., Hartmann, S., Okasha, S., Eds., *The European Philosophy of Science Association Proceedings Amsterdam 2009*, 303 – 313. Dordrecht: Springer.

Pigliucci, M., Boudry, M., Eds. (2013) *Philosophy of Pseudoscience.* Chicago/London: The University of Chicago Press.

Pinkava, J., Ed. (1987) *The Correspondence of the Czech Chemist František Wald with W. Ostwald, E. Mach, P. Duhem, J.W. Gibbs and Other Scientists of That Time.* Prague: Rozpravy Československé Véd – Sešit 5.

Planck, M. (1887a) Ueber das Princip der Vermehrung der Entropie. Erste Abhandlung. *Annalen der Physik* 266, 562 – 582.

Planck, M. (1887b) Ueber das Princip der Vermehrung der Entropie. Zweite Abhandlung. *Annalen der Physik* 267, 189 – 203.

Planck, M. (1887c) Ueber das Princip der Vermehrung der Entropie. Dritte Abhandlung. *Annalen der Physik* 268, 462 – 503.

Planck, M. (1891) Ueber das Princip der Vermehrung der Entropie. Vierte Abhandlung. *Annalen der Physik* 280, 385 – 428.

Planck, M. (1979) *Vorträge und Erinnerungen.* Darmstadt: Wissenschaftliche Buchgesellschaft.

Plath, P.J. (1990) Chemie. In Sandkühler, H.J., Ed., *Europäische Enzyklopädie zu Philosophie und Wissenschaften.* Hamburg: Meiner.

Preuss, H. (1960) Die gegenwärtige Situation der Quantenchemie. *Die Naturwissenschaften* 47, 241 – 249.

Preuss, H. (1962) *Grundriss der Quantenchemie.* Mannheim: Bibliographisches Institut.

Primas, H. (1983) *Chemistry, Quantum Mechanics and Reductionism.* Berlin: Springer.

Primas, H. (1985a) Kann Chemie auf Physik reduziert werden? Erster Teil: Das Molekulare Programm. *Chemie in unserer Zeit* 19, 109 – 119.

Primas, H. (1985b) Kann Chemie auf Physik reduziert werden? Zweiter Teil: Die Chemie der Makrowelt. *Chemie in unserer Zeit* 19, 160 – 166.

Primas, H. (1992) Umdenken in der Naturwissenschaft. *GAIA – Ecological Perspectives in Science, Humanities, and Economics* I, 5 – 15.

Primas, H., Müller-Herold, U. (1990) *Elementare Quantenchemie.* Stuttgart: B.G. Teubner.

Psarros, N. (1994) Sind die "Gesetze" der Konstanten und der Multiplen Proportionen Empirische Naturgesetze oder Normen? In Janich, P., Ed., *Philosophische Perspektiven der Chemie*, 53 – 63. Mannheim: BI Wissenschaftsverlag.

Psarros, N. (1999) *Die Chemie und ihre Methoden.* Weinheim: Wiley-VCH.

Psarros, N., Ruthenberg, K., Schummer, J., Eds. (1996) *Philosophie der Chemie – Bestandsaufnahme und Ausblick.* Würzburg: Königshausen & Neumann.

Putnam, H. (1975) *Mind, Language and Reality.* Vol. 2. Cambridge: Cambridge University Press.

Putnam, H. (1999) *Repräsentation und Realität.* Frankfurt am Main: Suhrkamp.

Quine, W.V.O. (1969) Ontological Relativity and Other Essays. New York: Columbia University Press.

Ramberg, P. (2000) The Death of Vitalism and the Birth of Organic Chemistry: Wöhler's Urea Synthesis and the Disciplinary Identity of Organic Chemistry. *Ambix* 47, 170 – 195.

Ramsay, W. (1895) Helium, a Gaseous Constituent of Certain Minerals, Part I. *Proceedings of the Royal Society of London* 58, 80 – 89.

Rancke Madsen, E. (1958) *The Development of Titrimetric Analysis till 1806.* Copenhagen: GEC GAD Publishers.

Regnault, H.V. (1851) *Kurzes Lehrbuch der Chemie* (French original: Paris 1847-1849). Braunschweig: Vieweg.

Restrepo, G. (2022) Chemical space: limits, evolution and modelling of an object bigger than our universal library. *Digital Discovery* 1, 568-585.

Richter, J.B. (1792/1968) *Anfangsgründe der Stöchiometrie.* Vol. 1. Hildesheim: Georg Olms (Reprint; Original Breslau & Hirschberg: Johann Friedrich Korn).

Richter, K., Laitko, H. (1962) Zur Gegenstandsbestimmung der Chemie. *Deutsche Zeitschrift für Philosophie* 10, 1278 – 1293.

Rickert, H. (1888) *Zur Lehre von der Definition.* Dissertation, Kaiser-Wilhelm-Universität Strassburg.

Riedel, E. (2004) *Allgemeine und Anorganische Chemie.* 8th edition. Berlin/New York: De Gruyter.
Rocke, A. (1984) *Chemical Atomism in the Nineteenth Century – From Dalton to Cannizzaro.* Columbus: Ohio State University Press.
Römpp (1975) Chemie-Lexikon. 7th ed. Hrsg. Neumüller, O.-A. *Band 1.* Stuttgart: Franckh'sche Verlagshandlung.
Rosenfeld, S., Bhushan, N. (2000) Chemical Synthesis – Complexity, Similarity, Natural Kinds, and the Evolution of a "Logic". In Bhushan, N., Rosenfeld, S., Eds., *Of Minds and Molecules – New Philosophical Perspectives on Chemistry*, 187–207. Oxford: Oxford University Press.
Roth, K. (2005) Eine unendliche chemische Geschichte. *Chemie in unserer Zeit* 39, 212–217.
Rowland, F.S. (1996) Der Abbau des stratosphärischen Ozons durch Fluorchlorkohlenwasserstoffe (Nobel-Vortrag). *Angewandte Chemie* 108, 1908–1921.
Russell, C.A. (1971) *The History of Valency.* Leicester: Leicester University Press.
Russell, C.A. (1987) The Changing Role of Synthesis in Organic Chemistry. *Ambix* 34, 169–180.
Ruthenberg, K. (1992) Was kann Timaios erklären? Wissenschaftstheoretische Anmerkungen zur platonischen Chemie. *Antike Naturwissenschaft und ihre Rezeption* 1/2, 44–58.
Ruthenberg, K. (1993) Vorgeschichte der Chemie: Ein Vergleich der theoretischen Konzeptionen von Platon und Aristoteles. *Antike Naturwissenschaft und ihre Rezeption* 3, 61–68.
Ruthenberg, K. (1994) Die Schwierigkeiten mit der Definition der Chemie. *Chemie in Labor und Biotechnik* 45(6), 303–306.
Ruthenberg, K. (1995a) Philosophische Probleme der Chemie – Enzyklopädische Artikel im Vergleich. *Praxis der Naturwissenschaften/Chemie* 44(4), 39–42.
Ruthenberg, K. (1995b) Das Verständnis von Auslösung und Beschleunigung chemischer Vorgänge im 19. Jahrhundert. In Schäfer, L., Ströker, E., Eds., *Naturauffassungen in Philosophie, Wissenschaft, Technik.* Vol. 3: *Aufklärung und späte Neuzeit*, 41–69. Freiburg/Munich: Karl Alber.
Ruthenberg, K. (1996a) Warum ist die Chemie ein Stiefkind der Philosophie? In Psarros, N., Ruthenberg, K., Schummer, J., Eds., *Philosophie der Chemie – Bestandsaufnahme und Ausblick* , 27–35. Würzburg: Königshausen & Neumann.
Ruthenberg, K. (1996b) Mathematik in und Mathematisierung der Chemie. In Janich, P., Psarros N., Eds., *Die Sprache der Chemie*, 101–112. Würzburg: Königshausen & Neumann
Ruthenberg, K. (1996c) Alwin Mittasch (1869–1953). *HYLE – International Journal for Philosophy of Chemistry* 2, 12.
Ruthenberg, K. (1997) Friedrich Adolf Paneth (short biography). *HYLE – International Journal for Philosophy of Chemistry* 3, 103–106.
Ruthenberg, K. (2005) Empedokles' Naturdichtung aus chemischer Perspektive. *Antike Naturwissenschaft und ihre Rezeption* 15, 7–12.
Ruthenberg, K. (2007a) "Frantisek Wald" (short biography). *HYLE – International Journal for Philosophy of Chemistry* 13, 55–61.
Ruthenberg, K. (2007b) Saving Chemical Phenomena outside the Scientific Community (review of F. Wald: Chemie Fasi). *Foundations of Chemistry* 9, 317–320.
Ruthenberg, K. (2008a) The "Stuffiness" of Ions – Ostwald as Anti-Atomistic Ionist. In Bertomeu-Sanchez, J.R., Burns, D.T., Van Tiggelen, B., Eds., *Neighbours and Territories.The Evolving Identity of Chemistry*, 403–410. Louvain-la-neuve: Memosciences.
Ruthenberg, K. (2008b) Chemistry without Atoms: Frantisek Wald. In Ruthenberg, K., van Brakel, J., Eds., *Stuff – The Nature of Chemical Substances*, 54–69. Würzburg: Königshausen & Neumann.
Ruthenberg, K., Ed. (2009a) *Frantisek Wald Essays 1891–1929.* Prague: Wald Press.

Ruthenberg, K. (2009b) Paneth, Kant, and the Philosophy of Chemistry. *Foundations of Chemistry* 11, 79–91.
Ruthenberg, K. (2010) Das Kant'sche Echo in Paneths Philosophie der Chemie. *Kant-Studien* 101, 465–479.
Ruthenberg, K. (2011) Chemietheorie "nach dem Vorgange von Mach". František Walds phänomenalistischer Ansatz in den "Annalen der Naturphilosophie". *Abhandlungen der Sächsischen Akademie der Wissenschaften zu Leipzig – Philologisch-historische Klasse* 82, 94–115.
Ruthenberg, K. (2012a) Frantisek Wald (1861–1930). In Hendry, R.F. et al., Eds., *Handbook of the Philosophy of Science.* Vol. 6: *Philosophy of Chemistry,* 125–131. Amsterdam: Elsevier.
Ruthenberg, K. (2012b) What Is Water? – Some Philosophical Considerations. In Hahn, H.P., Cless, K., Soentgen, J., Eds., *People at the Well*, 65–78. Frankfurt/New York: Campus.
Ruthenberg, K. (2015a) Als das Unmögliche noch möglich erschien. *Nachrichten aus der Chemie* 63, 639–643.
Ruthenberg, K. (2015b) Radicals, Reactions, Realism. In Scerri, E., McIntyre L., Eds., *Philosophy of Chemistry*, 183–199. Dordrecht: Springer Science+Business.
Ruthenberg, K. (2016a) Matter and Stuff: Two Sides of the Same Medal? In Le Moli, A., Cicatello, A., Eds., *Contemporary Lines.* Vol. 2 of *Understanding Matter*, 153–168. Palermo: New Digital Press.
Ruthenberg, K. (2016b) About the Futile Dream of an Entirely Riskless and Fully Effective Remedy: Thalidomide. *Hyle-International Journal for Philosophy of Chemistry* 22, 55–77.
Ruthenberg, K. (2019) Zur ontologischen Unterbestimmtheit von chemischen Stoffen. In Schwinger, E., Ed., *Das Anthropozän im Diskurs der Fachdisziplinen*, 83–95. Marburg: Metropolis.
Ruthenberg, K., Ed. (2020a) *Säuren – Eine kleine Kulturgeschichte.* Coburg: Edition Aumann.
Ruthenberg, K. (2020b) "Caught in the Amber": A Sketch of Chemical Underdetermination. In LaCaze, A., Osimani, B., Eds., *Uncertainty in Pharmacology – Epistemology, Methods, and Decisions*, 173–184. Cham: Springer Nature.
Ruthenberg, K. (2020c) Making Elements. In Scerri, E., Ghibaudi, E., Eds., *What Is a Chemical Element?*, 204–224. Oxford: Oxford University Press.
Ruthenberg, K. (2022) *Chemiephilosophie.* Berlin/Boston: De Gruyter.
Ruthenberg, K. (2023) Warum Lowry? *Nachrichten aus der Chemie* 71, 13–15.
Ruthenberg, K. (2024a). The Philosophical Neglect of Chemistry Revisited. In Ave Mets et al. (eds.) *Practical Realist Philosophy of Science*, 189–202.
Ruthenberg, K. (2024b) Bifurcations. *Foundations of Chemistry*, in press.
Ruthenberg, K., Chang, H. (2017) Acidity: Modes of Characterization and Quantification. *Studies in History and Philosophy of Science* 65–66, 121–131.
Ruthenberg, K., Chang, H. (2020) Glass and Life: The Biochemical Origins of pH. *Mitteilungen* 26, 63–87.
Ruthenberg, K., Gerontas, A. (2018) Säuren in der Antike und frühen Neuzeit. *Antike Naturwissenschaft und ihre Rezeption* 28, 227–242.
Ruthenberg, K., Harré, R. (2012) Philosophy of Chemistry as Intercultural Philosophy: Jaap van Brakel. *Foundations of Chemistry* 14, 193–203.
Ruthenberg, K., Kessel, N. (2020) Responsabilité et sous-détermination. In Bensaude-Vincent, B., Eastes, R.-E., Eds., *Philosophie de la Chimie*, 337–341. Louvain-la Neuve: deboeck Supérieur.
Ruthenberg, K., Martínez Gonzalez, J.C. (2017) Electronegativity and Its Multiple Faces: Persistence and Measurement. *Foundations of Chemistry* 19, 61–75.
Ruthenberg, K., Mets, A. (2020) Chemistry Is Pluralistic. *Foundations of Chemistry* 22, 403–419.

Ruthenberg, K., Psarros, N. (1994) František Wald und die phänomenologische Chemie. *Mitteilungen der Fachgruppe Geschichte der Chemie* 10, 17 – 30.
Ruthenberg, K., van Brakel, J., Eds. (2008) *Stuff – The Nature of Chemical Substances.* Würzburg: Königshausen & Neumann.
Rutherford, E. (1919) Collision of α Particles with Light Atoms III. Nitrogen and Oxygen Atoms. *Philosophical Magazine* 37, 571 – 580.
Scerri, E.R. (2007) The Periodic Table – Its Story and Its Significance. Oxford: Oxford University Press.
Scerri, E.R. (2016) *A Tale of Seven Scientists – and a New Philosophy of Science.* Oxford: Oxford University Press.
Scerri, E.R. (2020) On Chemical Natural Kinds. *Journal for General Philosophy of Science* 51, 427 – 445.
Scerri, E., McIntyre, L., Eds. (2015) *Philosophy of Chemistry – Growth of a New Discipline.* Dordrecht: Springer.
Scerri, E., Ghibaudi, E., Eds. (2020) *What Is a Chemical Element?* Oxford: Oxford University Press.
Scheringer, M. (1999) *Persistenz und Reichweite von Umweltchemikalien.* Weinheim: Wiley-VCH.
Schmidt, M. (1967) *Anorganische Chemie*, Band 1. Mannheim: BI Wissenschaftsverlag.
Schönbein, C.F. (1844/1996) *Kleine Abhandlungen.* Ed. H.-W. Schütt. Hildesheim: Olms-Weidmann.
Schorlemmer, C. (1889/1979) *Ursprung und Entwicklung der organischen Chemie* (Ostwalds Klassiker der exakten Wissenschaften 259). Leipzig: Geest & Portig.
Schummer, J. (1994) Die Rolle des Experiments in der Chemie. In Janich, P., Ed., *Philosophische Perspektiven der Chemie*, 27 – 51. Mannheim: BI Wissenschaftsverlag.
Schummer, J. (1996a) Philosophie der Stoffe, Bestandsaufnahme und Ausblicke: Von der philosophischen Entstofflichung der Welt zur ökologischen Relevanz einer Philosophie der Stoffe. In Psarros, N., Ruthenberg, K., Schummer, J., Eds., *Philosophie der Chemie – Bestandsaufnahme und Ausblick*, 143 – 164. Würzburg: Königshausen & Neumann.
Schummer, J. (1996b) *Realismus und Chemie.* Würzburg: Königshausen & Neumann.
Schummer, J. (2002) The Impact of Instrumentation on Chemical Species Identity – From Chemical Substances to Molecular Species. In Morris, P., Ed., *From Classical to Modern Chemistry – The Instrumental Revolution*, 188 – 211. London: Royal Society of Chemistry & Science Museum London.
Schummer, J. (2015) The Methodological Pluralism of Chemistry and Its Philosophical Implications. In Scerri, E., McIntyre, L., Eds., *Philosophy of Chemistry*, 57 – 72. Dordrecht etc.: Springer.
Schummer, J. (2017) Philosophie der Chemie. In Lohse, S., Reydon, Th., Eds., *Grundriss Wissenschaftsphilosophie – Die Philosophien der Einzelwissenschaften*, 229 – 251. Hamburg: Meiner.
Schummer, J. (2020) The Chemical Prediction of Stratospheric Ozone Depletion: A Moral Model of Scientific Hazard Foresight. *Hyle – International Journal for Philosophy of Chemistry* 26, 31 – 54.
Schütt, H.W. (1992) *Eilhard Mitscherlich – Baumeister am Fundament der Chemie.* Munich: Oldenbourg.
Schwab, G.-M. (1931) *Katalyse – Vom Standpunkt der Chemischen Kinetik.* Berlin: Julius Springer.
Schwabe, K. (1953) *Fortschritte der pH-Messtechnik.* Berlin: VEB Verlag Technik.
Seifert, V. A. (2023) *Chemistry's Metaphysics.* Cambridge: Cambridge University Press.
Servos, J.W. (1990) *Physical Chemistry from Ostwald to Pauling.* Princeton NJ: Princeton University Press.
Simoes, A. (2006) In between Worlds: G.N. Lewis, the Shared Pair Bond and Its Multifarious Contexts. *Journal of Computational Chemistry* 28, 62 – 72.
Simon, R. (1980) *Chemie – Dialektik – Theorienentwicklung.* Berlin: Akademie-Verlag.
Simon, R., Niedersen, U., Kertscher, G. (1982) *Philosophische Probleme der Chemie.* Berlin: VEB Deutscher Verlag der Wissenschaften.

Sjöström, H., Nilsson, R. (1972) *Thalidomide and the Power of the Drug Companies.* Middlesex: Penguin (German translation as *Contergan oder die Macht der Arzneimittelkonzerne*, 1975, Berlin: VEB Verlag Volk und Gesundheit).

Soentgen, J. (1997) *Das Unscheinbare.* Berlin: Akademie-Verlag.

Soentgen, J. (2008) Stuff: A Phenomenological Definition. In Ruthenberg, K., van Brakel, J., Eds., *Stuff – The Nature of Chemical Substances*, 71 – 91. Würzburg: Königshausen & Neumann.

Soentgen, J. (2019) *Konfliktstoffe.* Munich: oekom.

Solomon, S., Ivy, D.J., Kinnison, D., Mills, M.J., Neely III, R.R., Schmidt, A. (2016) Emergence of Healing in the Antarctic Ozone Layer. *Science* 333, 269 – 274.

Staab, H.A. (1959) *Einführung in die theoretische organische Chemie.* Weinheim: Verlag Chemie.

Stahl, G.E. (1729) Chymia Rationalis et Experimentalis. Leipzig: Caspar Jacob Eyssel.

Stanford, P.K. (2006) *Exceeding Our Grasp.* Oxford: Oxford University Press.

Stanford, P.K., Kitcher, P. (2000) Refining the Causal Theory of Reference for Natural Kind Terms. *Philosophical Studies* 97, 99 – 129.

Stark, J. (1908) Zur Energetik und Chemie der Bandenspektra. *Physikalische Zeitschrift* 9, 85 – 94.

Stark, J. (1914) *Elektrische Spektralanalyse chemischer Atome.* Leipzig: S. Hirzel.

Stark, J. (1915) *Prinzipien der Atomdynamik. III. Teil: Die Elektrizität im chemischen Atom.* Leipzig: S. Hirzel.

Statham, G. (2017) The Manipulation of Chemical Reactions: Probing the Limits of Interventionism. *Synthese* 194, 4815 – 4838.

Steininger, B. (2017) Berührungswirkungen: Katalyse als Kontaktforschung. In Harrasser, K., Ed., *Auf Tuchfühlung: eine Wissensgeschichte des Tastsinns*, 25 – 37. Frankfurt am Main: Campus.

Stephan, A. (2007) *Emergenz – Von der Unvorhersagbarkeit zur Selbstorganisation.* Münster: Mentis.

Stephens, T., Brynner, R. (2001) *Dark Remedy.* Cambridge: Perseus.

Stock, J.T., Orna, M.V. (1986) *The History and Preservation of Chemical Instrumentation.* Dordrecht: D. Reidel.

Stranges, A.N. (1982) *Electrons and Valence.* College Station: Texas A&M University Press.

Ströker, E. (1967) *Denkwege der Chemie – Elemente ihrer Wissenschaftstheorie.* Freiburg/Munich: Karl Alber.

Ströker, E. (1968) Element und Verbindung. Zur Wissenschaftsgeschichte zweier chemischer Grundbegriffe. *Angewandte Chemie* 80, 747 – 753.

Ströker, E. (1982) *Theoriewandel in der Wissenschaftsgeschichte.* Frankfurt am Main: Vittorio Klostermann.

Strube, I. (1984) *Georg Ernst Stahl.* Leipzig: BSB B.G. Teubner.

Sun, Z., Tomanek, D. (1989) Cold Fusion: How Close Can Deuterium Atoms Come Inside Palladium? *Physical Review Letters* 63, 59 – 61.

Sykes, P. (1986) *A Guidebook to Mechanism in Organic Chemistry.* 6th edition. Singapore: Longman.

Szabadvary, F. (1966) *Geschichte der Analytischen Chemie.* Budapest: Akadémiai Kiadó.

Theobald, D.W. (1973) *Grundzüge der Wissenschaftsphilosophie.* Stuttgart: Philip Reclam jun.

Theobald, D.W. (1976) Some Considerations on the Philosophy of Chemistry. *Chemical Society Reviews* 5, 203 – 213.

Thiele, J. (1973) Franz Walds Kritik der theoretischen Chemie. *Annals of Science* 30, 417 – 433.

Thomson, J.J. (1913) *Rays of Positive Electricity and Their Application to Chemical Analyses.* London: Longmans, Green & Co.

Timmermans, J. (1940) *Chemical Species.* New York: Chemical Publishing.

Tugendhat, E., Wolf, U. (1983) Logisch-semantische Propädeutik. Stuttgart: Philipp Reclam jun.

van Brakel, J. (1986) The Chemistry of Substances and the Philosophy of Mass Terms. *Synthese* 69, 291 – 324.
van Brakel, J. (1992) Natural Kinds and Manifest Forms of Life. *Dialectica* 46, 243 – 261.
van Brakel, J. (1997) Chemistry as the Science of the Transformation of Substances. *Synthese* 111, 253 – 282.
van Brakel, J. (2000) *Philosophy of Chemistry.* Leuven: Leuven University Press.
van Brakel, J. (2005) On the Inventors of XYZ. *Foundations of Chemistry* 7, 57 – 84.
van Brakel, J. (2006) Kant's Legacy for the Philosophy of Chemistry. In Baird, D., Scerri, E., McIntyre, L., Eds., *Philosophy of Chemistry – Synthesis of a New Discipline*, 69 – 91. Dordrecht: Springer.
van Brakel, J. (2012) Substances: The Ontology of Chemistry. In Hendry, R.F., Needham, P., Woody, A., Eds., *Handbook of the Philosophy of Science.* Vol. 6: *Philosophy of Chemistry*, 191 – 229. Amsterdam: Elsevier.
van Brakel, J. (2014) Philosophy of Science and Philosophy of Chemistry. *HYLE – International Journal for Philosophy of Chemistry* 20, 11 – 57.
van Brakel, J. (2013) František Wald's Empiricism. *HYLE – International Journal for Philosophy of Chemistry* 19, 161 – 183.
van Brakel, J., Vermeeren, H. (1981) On the Philosophy of Chemistry. *Philosophy Research Archives* 7, 501 – 552.
van der Vet, P. (1979) The Debate between F.A. Paneth, G. von Hevesy, and K. Fajans on the Concept of Chemical Identity. *Janus* 66, 285 – 303.
van der Vet, P. (1987) *The Aborted Takeover of Chemistry by Physics.* Proefschrift (Dissertation), University of Amsterdam.
van der Waals, J.D. (1988) *On the Continuity of the Gaseous and Liquid States.* Ed. J.S. Rowlinson. Amsterdam: North-Holland.
van Fraassen, B.C. (2008) *Scientific Representation.* Oxford: Oxford University Press.
van't Hoff, J.H. (1884/1985) *Studien zur chemischen Dynamik* (Ostwalds Klassiker 265). Leipzig: Geest & Portig.
Vermeeren, H.P.W. (1986) Controversies and Existence Claims in Chemistry: The Theory of Resonance. *Synthese* 69, 273 – 290.
Verne, J. (1865/2008) *Von der Erde zum Mond.* Munich: Deutscher Taschenbuch Verlag.
Vihalemm, R. (2016) Chemistry and the Problem of Pluralism in Science: An Analysis Concerning Philosophical and Scientific Disagreements. *Foundations of Chemistry* 18, 91 – 102.
von Hartmann, E. (1889) *Das Grundproblem der Erkenntnistheorie.* Leipzig: Wilhelm Friedrich.
von Hartmann, E. (1914) *Kritische Grundlegung des transzendentalen Realismus.* 4th edition. Leipzig: Alfred Kröner.
von Hevesy, G., Paneth F. (1923) *Lehrbuch der Radioaktivität.* Leipzig: Johann Ambrosius Barth.
Wald, F. (1889) *Die Energie und ihre Entwertung.* Leipzig: Wilhelm Engelmann.
Wald, F. (1895) Die Genesis der stöchiometrischen Grundgesetze. *Zeitschrift für physikalische Chemie* 18, 337 – 375.
Wald, F. (1897a) Die Chemischen Proportionen I. *Zeitschrift für Physikalische Chemie* 22, 253 – 267.
Wald, F. (1897b) Elementare Chemische Betrachtungen. *Zeitschrift für Physikalische Chemie* 24, 633 – 650.
Wald, F. (1899) Was ist ein chemisches Individuum? *Zeitschrift für physikalische Chemie* 28, 13 – 16.
Wald, F. (1902a) Kritische Studie über die wichtigsten chemischen Grundbegriffe. *Annalen der Naturphilosophie* 1, 15 – 19.

Wald, F. (1902b) Kritische Studie über die wichtigsten chemischen Grundbegriffe. *Annalen der Naturphilosophie* 1, 182–216.
Wald, F. (1902c) Über einen alten Denkfehler in der Chemie. *Annalen der Naturphilosophie* 1, 470–472.
Wald, F. (1903) Über die Mannigfaltigkeit chemischer Erscheinungen. *Annalen der Naturphilosophie* 2, 108–132.
Wald, F. (1904) Neue Ableitung der Gibbsschen Phasenregel. *Annalen der Naturphilosophie* 3, 283–293.
Wald, F. (1906) Bausteine zu einer neuen chemischen Theorie. *Annalen der Naturphilosophie* 5, 271–290.
Wald, F. (1907a) Das nächste Problem der Chemie. *Annalen der Naturphilosophie* 6, 1–15.
Wald, F. (1907b) Nochmals über das nächste Problem der Chemie. *Annalen der Naturphilosophie* 6, 229–240.
Wald, F. (1909a) Mathematische Beschreibung chemischer Vorgänge. *Annalen der Naturphilosophie* 8, 214–265.
Wald, F. (1909b/2009) Chemie a mathematika (Chemistry and Mathematics). In Ruthenberg, K., Ed., *Frantisek Wald Essays 1891–1929* , 132–147. Prague: Wald Press.
Wald, F. (1918/2004) *Chemie Fasi.* Prague: Vydala Univerzita Karlova v Praze.
Walden, P. (1924) *Chemie der freien Radikale.* Leipzig: S. Hirzel.
Walden, P. (1928) Die Bedeutung der Wöhlerschen Harnstoff-Synthese. *Die Naturwissenschaften* 16, 835–849.
Walden, P. (1929) *Salts, Acids, and Bases/Electrolytes/Stereochemistry.* New York: McGraw-Hill.
Wegscheider, R. (1918) Über den Begriff der chemischen Elemente und über Atomgewichtstabellen. *Zeitschrift für Physikalische Chemie* 19, 741–749.
Weidenbruch, M. (1994) Das Elektronenoktett: Renaissance einer totgeglaubten Regel. *Chemkon* 1, 15–20.
Weyer, J. (2018a) *Geschichte der Chemie.* Vol. 1: *Altertum, Mittelalter, 16. bis 18. Jahrhundert.* Berlin: Springer Spektrum.
Weyer, J. (2018b) *Geschichte der Chemie.* Vol. 2: *19. und 20. Jahrhundert.* Berlin: Springer Spektrum.
Wheland, G.W. (1944) *The Theory of Resonance and Its Application to Organic Chemistry.* New York: John Wiley & Sons.
WMO (2018) *Scientific Assessment of Ozone Depletion – Global Ozone Research and Monitoring Project–Report No. 58.* Geneva.
Wöhler, F. (1828) Über künstliche Bildung des Harnstoffs. *Annalen der Chemie und Pharmazie* 12, 253–256.
Wolters, G. (1991) Vorwort zum Neudruck 1985. In Mach, E., *Analyse der Empfindungen.* Reprint of the 9th edition. Darmstadt: Wissenschaftliche Buchgesellschaft, IX-XXIII.

List of Figures

https://doi.org/10.1515/9783111401041-016

Name Index

https://doi.org/10.1515/9783111401041-017

Subject Index

https://doi.org/10.1515/9783111401041-018

www.ingramcontent.com/pod-product-compliance
Lightning Source LLC
Chambersburg PA
CBHW060738240325
23923CB00056B/189